AF587525

VoIP Performance of the Relay-enhanced IEEE 802.16m Wireless Broadband System

Karsten Klagges

AACHENER BEITRÄGE ZUR MOBIL- UND TELEKOMMUNIKATION

Herausgeber:
Universitätsprofessor Dr.-Ing. Bernhard Walke

Bibliografische Information Der Deutschen Bibliothek
Die Deutsche Bibliothek verzeichnet diese Publikation in der Deutschen Nationalbibliografie; detaillierte bibliografische Daten sind im Internet über http://dnb.ddb.de abrufbar.

1. Auflage 2015
ISBN: 978-3-95886-022-3
Aachener Beiträge zur Mobil- und Telekommunikation; Band 73

Wissenschaftsverlag Mainz
Süsterfeldstr. 83, 52072 Aachen
Telefon: 02 41 / 2 39 48 oder 02 41 / 87 34 34
Fax: 02 41 / 87 55 77
www.verlag-mainz.de

Herstellung: Druckerei Mainz GmbH,
Süsterfeldstr. 83, 52072 Aachen
Telefon 02 41 / 87 34 34; Fax: 02 41 / 87 55 77

Gedruckt auf chlorfrei gebleichtem Papier

D 82 (Diss. RWTH Aachen University, 2014)

ABSTRACT

In 2007 the International Telecommunication Union, Radiocommunication Sector (ITU-R) published evaluation guidelines for future mobile broadband radio networks including a minimal VoIP capacity that the IEEE 802.16m WiMAX system had to meet, proven by system level simulation. These evaluation guidelines are the foundation of this work that shows that relay-enhanced 4G networks do not only meet the requirements but exceed them by far. The key-requirement for packet based VoIP services that has been set in the evaluation guidelines demands that the end-to-end packet delay of at least 98% of user data traffic stays below 50 ms.

Since 2011 proposals of ComNets to implement relays into mobile radio networks to increase system capacity are part of all 4G systems (LTE and WiMAX). This work investigates the voice over IP (VoIP) capacity of the relay-enhanced WiMAX system by system simulation. The work presents a modular approach that has been developed by my colleagues and myself to implement the WiMAX protocol stack, also known as WiMAC, into the openWNS simulator. The approach is based on atomic protocol functions that are bound to so called FU and interconnected in a network of functional units (FUs) and thereby implement the WiMAX protocol bit-by-bit. Besides the performance evaluation of VoIP services based on the ITU-R scenarios and channel model, this work also determines the optimal radio resource configuration for up- and downlink and furthermore investigates the impact of RS location on system capacity. Core functions of the simulator are being validated by comparison of equivalent results of other simulators.

Cumulative distribution functions of the end-to-end delay of VoIP user data packets, determined by simulation show that relays double the carried VoIP traffic load compared to a scenario without relay stations (RSs) due to the improved prediction of the uplink channel state that is the bottleneck of the system.

The performance enhancement is a surprise since relays increase the end-to-end packet delay and packet error due to an additional transmission compared to conventional single-hop operation.

KURZFASSUNG

Die ITU-R hat in 2007 Richtlinien für die Bewertung zukünftiger Breitband Mobilfunksysteme erlassen und den Nachweis durch Systemsimulation gefordert, dass IEEE 802.16m-WiMAX eine bestimmte VoIP Kapazität hat. Diese Richtlinien werden in dieser Arbeit als Grundlage genommen und der Nachweis geführt, dass durch Relais verstärkte mobile Breitbandsysteme geeignet sind, weil mehr als den von 4G Systemen geforderten VoIP Verkehr zu tragen.

VoIP Pakete dürfen maximal 50 ms verspätet sein und höchstens 2% der Pakete dürfen ausfallen.

Seit 2011 sind die von ComNets zur Leistungssteigerung von Breitband-Mobilfunksystemen vorgeschlagenen Relais Bestandteil aller Standards für 4G Systeme (LTE und WiMAX).

Diese Arbeit untersucht die VoIP Kapazität von WiMAX Systemen mit Relais durch Systemsimulation. Dabei wird ein von mir und meinen Kollegen bei ComNets entwickeltes Software Entwurfsverfahren zur Realisierung des WiMAC genannten Protokollstapels für die Systemsimulation in der Umgebung des openWNS Simulators eingesetzt. Das Verfahren beruht auf atomaren Protokollfunktionen, die als sog. Functional Units zu Netzen von FUs zusammengesetzt werden und dabei das WiMAX Protokoll bitgenau implementieren. Neben der Leistungsbewertung des WiMAX Mobilfunksystems in einer von der ITU-R spezifizierten Betriebsumgebung und Benutzung der vorgegebenen Kanalmodelle werden die bestmögliche Konfiguration der WiMAX Funkbetriebsmittel für Abwärts- bzw. Aufwärtsstrecke und der Einfluss der Standorte von Relais auf die VoIP Kapazität ermittelt. Grundfunktionen des Simulators werden durch Vergleich mit Ergebnissen anderer Simulatoren validiert.

Anhand der durch Systemsimulation ermittelten Verzögerungsverteilungsfunktion von VoIP Paketen kann bei bester Konfiguration mit Relais der tragbare VoIP Verkehr gegenüber einem System ohne Relais fast verdoppeln werden. Es wird nachgewiesen, dass das an der in Relais verstärkten Systemen erzielbaren besseren Prädiktion der Kanalqualität an der Aufwärtsstrecke liegt, die in WiMAX Systemen ein deutlicher Engpass ist.

Die mit Relais erzielbare Kapazitätssteigerung ist überraschend, weil Relais die Verzögerung von VoIP Paketen durch multihop Kommunikation vergrößern und die Übertragung von VoIP Paketen zwischen Basisstation und Mobilstation über ein Relais eine gegenüber einhop Kommunikation erhöhte Fehleranfälligkeit erwarten lassen.

Contents

CHAPTER 1

Introduction

Contents

1.1 Software Platforms for modular Protocol Stacks

Protocol simulators emerged to be an essential tool for performance evaluation of recent 4G mobile radio systems. Besides link level simulators that only model the radio transmission over a point-to-point channel they serve as important tools for performance evaluation in terms of system throughput capacity. In addition system level simulators offer means to assess service quality for time critical services like streaming or interactive services. With the decision by 3rd Generation Partnership Project (3GPP) standardization group not to offer any more service based on circuit switched technology in 4G systems, realtime oriented services like voice service and video streaming will be based on packet switching. Therefore it is especially of interest to evaluate the performance of packet based realtime services as investigated for the VoIP service in this thesis.

With the help of a system level simulator, the protocol designer is able to evaluate various protocol settings quickly without the need to perform measurements in hardware based testbeds. The size of the system is limited only by the performance of the central processing unit (CPU) and the memory size of the execution host. Commercially available personal computers have sufficient processing power to simulate large scenarios comprising many base stations (BSs) in detail. This allows the protocol designer to reduce the development expense substantially.

Modular protocol stacks, developed on a simulator platform may also speed up the process of rapid prototyping of a real system implementation. The protocol configuration used in a simulator can be copied onto a prototype thereby specifying its protocol software.

1.2 Contribution of this Thesis

Functional units are the basis of the modular software architecture in openWNS. With non-ambiguous defined interfaces, FUs assure the compatibility of protocol components and their functions. A network of multiple connected FUs realizes the function of a complete protocol layer as a functional unit network (FUN). This work presents the extensions that have been made to enable the FUN to realize the Institute of Electrical and Electronics Engineers (IEEE) 802.16m system in the simulator.

The radio resource scheduler of the WiMAX media access control protocol (WiMAC) simulator module that has been developed as part of this work follows a strict abstraction of logical resource units and physical resource blocks. This enables the resource scheduler to be supplied in protocol stacks for any protocol that supports discrete radio resource allocation in a periodically recurring radio frame.

This work includes the performance evaluation of the relay-enhanced IEEE 802.16m system under VoIP traffic load and compares the system capacity with the BS-only scenario. The evaluation is based on the WiMAC module which makes use of the generic components provided by the openWNS simulator platform. It evaluates to what degree time-critical services like VoIP are served in the relay-enhanced system taking quality of service (QoS) requirements into account.

In summary this work establishes that

- the IEEE 802.16m system can be implemented entirely with the help of FUs that are modular and reusable
- a generic radio resource scheduler can be realized that is able to operate for both, time division multiplex (TDM) and frequency division multiplexing (FDM) with interchangeable disciplines
- IEEE 802.16m without decode and forward (DF) relays meets the IMT-Advanced (IMT-A) system family requirements for VoIP defined by the ITU-R
- IEEE 802.16m with optimally located DF relays outperform the IMT-A requirements for VoIP by far
- optimally placed relays that maximize downlink (DL) capacity also maximize uplink (UL) capacity
- minor displacement of relay positions from optimum has only minor impact on VoIP performance

1.3 Outline

Chapter 2 gives a brief introduction into fundamental principles of cellular radio networks. Chapter 3 presents the most important parts of the IEEE 802.16 protocol that are relevant for this work. Chapter 4 presents the models and scenarios that have been implemented in the simulator. Chapter 5 presents the concepts of the modular components of the protocol stacks to model the IEEE 802.16 protocol by means of the openWNS simualtor platform. Chapter 6 shows performance evaluation results for the VoIP serbvice provided by the IEEE 802.16 system represented by the openWNS simulator. Chapter 7 summarizes and concludes the work.

CHAPTER 2

Cellular Radio Networks and Relay Stations

Contents

2.1 Cellular Radio Networks

The major goal of cellular radio networks is to cover a wide service area with multiple BSs each covering a limited service area called a cell where its mobile stations (MSs) are roaming around. Cells served by BS overlap in part and usually are represented by contiguous hexagons each served by a BS. User terminals (MSs) associate to that BS they receive the strongest signal from (MacDonald, 1979).

Operators of commercial cellular networks need a license from government for the frequency bands used. Licenses fees in tendency are very high and therefore an operator aims to serve as many users as possible per frequency band unit. Signal pathloss attenuates radio signal strength and therefore the range of a BS expressed in the cell radius is dependent on the transmit power and is limited in general, since not only the BS needs to reach its MSs but MSs that have limited battery capacity need to reach the BS. With multiple base station sites the network can reuse radio channels in sufficiently distant cells where the co-channel interference of two cells operating on the same radio channel is below a certain limit.

The capacity of a radio link connecting two stations is limited by the signal to interference plus noise ratio (SINR) at the receiver. SINR γ at the receiver is given by Eq. (2.1) where P_C is the carrier power at the receiver, P_{In} is the interference power caused by interferer n at the receiver and N_0 the thermal noise power.

$$\gamma = \frac{P_C}{\sum\limits_{n \in N} P_{In} + N_0} \tag{2.1}$$

The receiver of a MS served by a BS on the so-called DL in a cellular system is interfered by downlink signals originating from (typically at least six) co-channel BSs serving their MSs. The receiver at the BS not only receives on UL signals of a MS roaming in its cell but also interference power from MSs transmitting uplink in co-channel cells.

Under adaptive modulation and coding, γ linearly translates into the throughput of the radio link. To maximize the throughput γ needs to be increased. A higher transmit power applied at all BSs would result in both, a higher received carrier strength and increased co-channel interference. Hence, increasing transmit power in general is not the solution to maximize link throughput and thereby cell capacity. The following presents known techniques for improving cell capacity in cellular networks by interference reduction.

2.1.1 Clustering

Cellular radio systems usually subdivide its licensed frequency spectrum into a set of frequency channels. Cells are aggregated into clusters of cells. Within a cluster, a frequency channel group is used only once in one of the cluster's cells. The number of cells each owing an exclusive group of frequency channels in a cluster is called the cluster order N. In a regular grid of hexagonal cells the possible cluster orders are given by Eq. (2.2).

$$N = i^2 + ij + j^2 \quad i, j \in \mathbb{N} \tag{2.2}$$

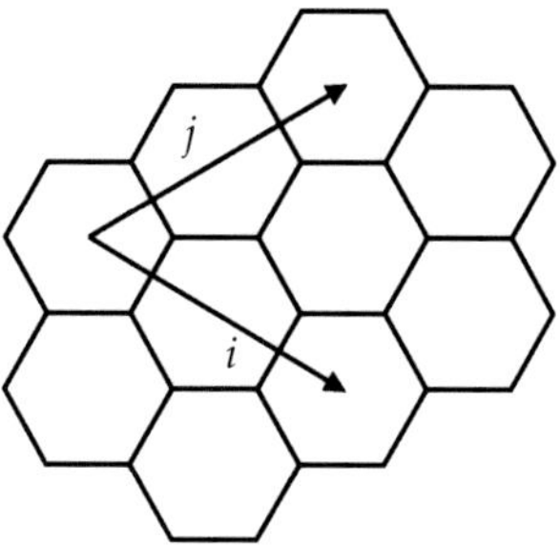

Figure 2.1: Hexagonal Coordinate System

i and j are the indices in the hexagonal coordinate system. Figure 2.1 shows how the hexagonal coordinate system aligns the unit vectors.

Figure 2.2 shows two cell clusters of order three that can be created with $i = 1$ and $j = 1$. Figure 2.3 shows two cell clusters of order four that can be constructed with $i = 0$ and $j = 2$. The numbers denote the frequency channel group used in the cell. Clustering cells results in a re-use distance d and results

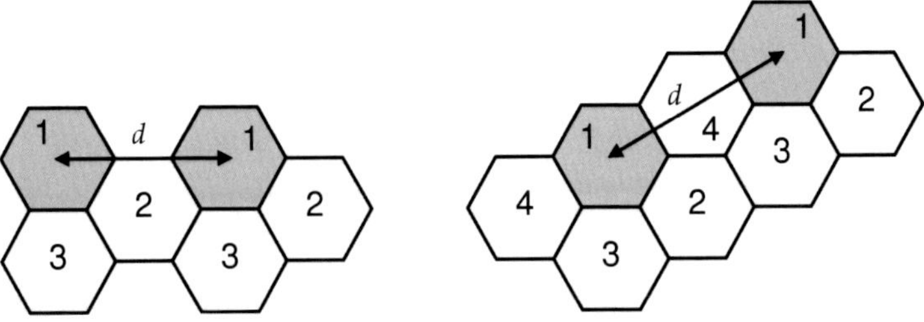

Figure 2.2: Cluster order 3

Figure 2.3: Cluster order 4

in lower interference power at the receivers for both, DL and UL transmissions the larger d is. Re-use distance d with a cell radius R is given in Eq. (2.3) for equally sized hexagonal cells.

$$d = \sqrt{3N} \cdot R \tag{2.3}$$

Clustering is a commonly used technology in cellular radio and allows to keep a sufficient low mean interference level by static radio network planning.

2.1.2 Fractional Frequency Reuse

To achieve higher spectral efficiency J. Halpern (Halpern, 1983) introduced fractional frequency reuse for cellular radio networks. Unlike with static clustering of cells, explained in Section 2.1.1, parts of the cell are either served by a group of commonly shared frequency channels or by a channel group derived from one of a cluster order. MSs that do not suffer from heavy interference power due to their location in a cell may re-use the radio channels assigned to direct neighbor cells.

Figure 2.4 shows a cellular system that re-uses a shared frequency channel group number 1 according to cluster order one for MSs located close to a BS. Cell-edge regions highlighted by hatching are served by channel groups following re-use-3

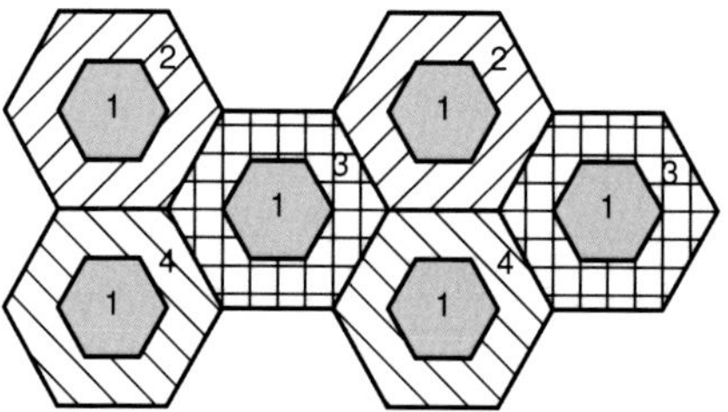

Figure 2.4: Fractional Frequency Reuse

according to cluster order three namely channel groups 2, 3 and 4. In total the system frequency bandwidth is divided into four channel groups 1 to 4 of which one may have a number of channels different from that of the three other groups. As shown in (Halpern, 1983), the benefit of fractional frequency re-use is an increased cell spectral efficiency. Moreover, flexibility of cellular system design is higher since,under fractional frequency re-use the number of channels grouped according to the cluster order can be chosen appropriately. E.g. the re-use-1 region of each cell grow or shrink according to the radio propagation conditions.

2.1.3 Sectors

With the help of a sector antenna, a BS can cover only a part of a cell while other parts are covered by different sector antennas. With the commonly used 120 ° sector antennas a BS serves three sectors independently of each other. Although the signals of the sector antennas are not completely separated but partly overlapping at the sector edges, dividing a cell into sectors can either increase the spectral efficiency of a cellular network significantly or reduce the cluster order and maintain the spectral efficiency (Chan, 1992; MacDonald, 1979; Wacker, Laiho-Steffens, Sipila, & Heiska, 1999).

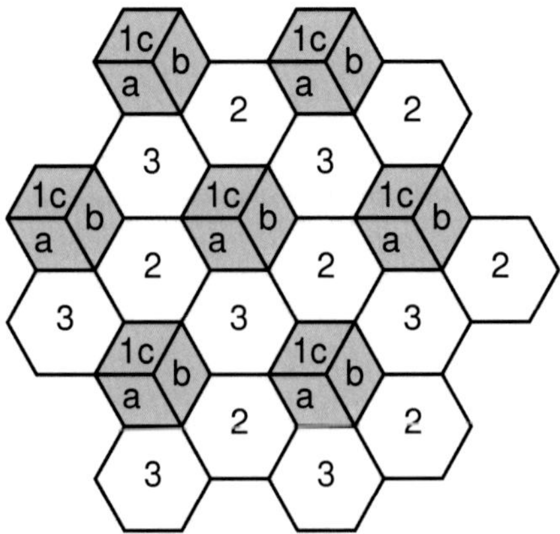

Figure 2.5: Sectoring

Figure 2.5 shows a cellular network with cluster order three and three sectors per cell. Sectors of cells with frequency channel groups 2 and 3 are not shown for better clarity.

2.1.4 Interference Control

Interference control strategies actively take decisions to avoid co-channel interference of neighbor cells. Interference control schemes can be classified by their time-scale of operation.

Static interference control is maintained at the time of cellular network planning. There, radio resources like frequency channel or time slot are statically assigned to a BS and are allocated to immediate neighbor cells according to the cluster order chosen. Static interference control operates on the time scale of days or more.

Semi-static interference control re-configures dedicated radio resources on a time scale of down to seconds. A BS may then shift unused radio resources to another BS which traffic load exceeds the available resources (TSG-RAN WG1, 2006).

Dynamic interference control also known as interference avoidance works on a time scale of a few radio frames, say milliseconds. It reacts almost instantly on changing radio resource demands in a cell and requires fast communication links between neighbor BSs to communicate resource allocations planned in the near future between cells (Boudreau et al., 2009).

Interference control strategies can further be classified by the range of the interference control. *Global control* requires an omniscient entity that has knowledge about the cell specific traffic load and the mutual interference of all BSs to be able to allocate radio resource interference free to BSs in the network. *Decentralized control* is based on resource allocation across neighbor cells relying on a direct communication between BSs. The performance of interference control schemes is investigated in (Necker, 2008; Wolz, Muehleisen, Klagges, & Einhaus, 2010).

Local schemes are typically for current 2G/3G systems and do not use any information exchange between neighbor BSs but estimate resource allocation of neighbor cells by mutual observation. If a cellular system is uniformly loaded, then interference control based resource allocation can reach a steady state and improve cell throughput (Chang, Tao, Zhang, & Kuo, 2013).

2.2 Relaying

ITU-R specified IMT-Advanced mobile radio systems are foreseen to be deployed in diverse geographic regions, like urban, suburban or rural areas. ComNets researchers (Pabst et al., 2004) have shown that decode-and-forward RSs are a cost-effective solution to improve SINR for highly populated areas with heavy shadowing and to provide coverage to rural areas.

The purpose of relays in a cell is either to improve radio coverage of a cell or to increase throughput capacity (Esseling,

Walke, & Pabst, 2004). In relay enhanced cellular IEEE 802.16m systems, the conventional BS is replaced by a multi-hop relay base station (MR-BS) that supports RS specific service and one or more RSs. In a DL communication cycle, at first the BS transmits a packet intended for a remote MS to the respective RS. In a second step the RS transmits the packet to the remote MS. For UL traffic the remote MS transmits its packet to the RS as soon as the RS has granted UL radio resources to the MS. Afterwards the RS transmits the packet to the BS. Recent investigations have shown that RSs may significantly improve cell coverage and cell capacity (Esseling, Pabst, & Walke, 2005; Genc, Murphy, & Murphy, 2008; Kusuda, Yamamoto, & Yoshida, 2005; Sambale & Walke, 2012a; Voudouris et al., 2012).

2.2.1 Motivation

It turned out that RSs can be employed to either improve the cell range to better cover large-area scenarios or to improve capacity of cells in densely populated areas.

2.2.1.1 Coverage Extension

Deploying a cellular network with full radio coverage to all locations in the cells is a challenging task and usually requires high capital expenditure (CAPEX). In sparsely populated regions the deployment of a conventional cellular network without relays may not be profitable. To reduce the initial network deployment costs, relays may be a cost-effective option. In such a case the relay stations are placed close to the border of a conventional cell. User terminals that are not in the service range of the base station then can be served by the relay station. Since the base station and the relay station transmit the packet twice across a two-hop link coverage extension with relay stations reduces cell spectral efficiency compared to a single-hop cell. Coverage extension by relays may also be considered as a measure to save energy ans support so-called green communications.

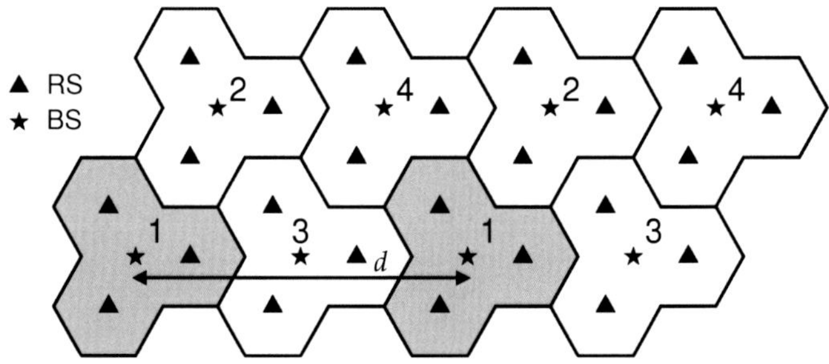

Figure 2.6: Cluster order 4 with Relays

In Fig. 2.6 three RSs are placed at the border of three-sector cells serving to extend the range of the BS by RSs. If RSs are assumed to operate the same transmission power as the BSs the cell area covered by a BS is extended to up to 300%. Consequently the re-use distance increases by a factor of $\sqrt{3}$, see Eq. (2.4).

$$d = 3\sqrt{N} \cdot R \tag{2.4}$$

2.2.1.2 Throughput Capacity Enhancement

The performance evaluation in Chapter 6 is focusing scenarios aiming at the throughput capacity enhancement. In dense populated areas, shadowing by buildings and indoor usage of radio services are reasons for low signal strength at MSs. Buildings and indoor usage prevents line of sight (LoS) radio propagation conditions between BS and MS. Placing RSs at street crossings and/or at locations for serving badly covered areas may substantially increase the probability of a MS to find a broadcast control channel of MR-BS or RS with good channel conditions (high SINR). Consequently, network capacity and spectral efficiency are increased since all MSs are served with better SINR compared to a conventional cell.

Improving system performance with relays is possible if relay stations are placed properly, namely served by line-of-sight links from the BS to maintain a high-capacity so-called backhaul link.

2.2.2 Relay and Base Station Operation in Time Division Multiplex

Integration of relays into a cellular radio network - in the preferred way of relay operation - requires orthogonal radio resources allocated to both, the BS for serving MS on the access link and RS on the backhaul link and for RSs for serving their MSs on the access link. Orthogonality is established by TDM mode of operation. The system needs to assure that RSs can operate unaffected by the operation of the BS in the cell. Resources for relay operation may be provided either in time or frequency domain. Following standard IEEE 802.16m the two designs of integration of RSs into radio frames of 4th generation (4G) cellular systems are explained assuming in the examples shown one RS per cell only.

2.2.2.1 Time Division Duplex

Under TDM, BS and RS share the same frequency channel. The system reserves radio resources for RSs and the BS in alternating phases of a periodic radio frame.

Figure 2.7 shows a simplified radio frame that is divided into a DL and a UL sub-frame. Each sub-frame contains radio resources dedicated either to DL transmission of a BS or RS or UL transmission of MSs to their serving BS or RS in time division duplex (TDD) mode of operation. Guard periods are

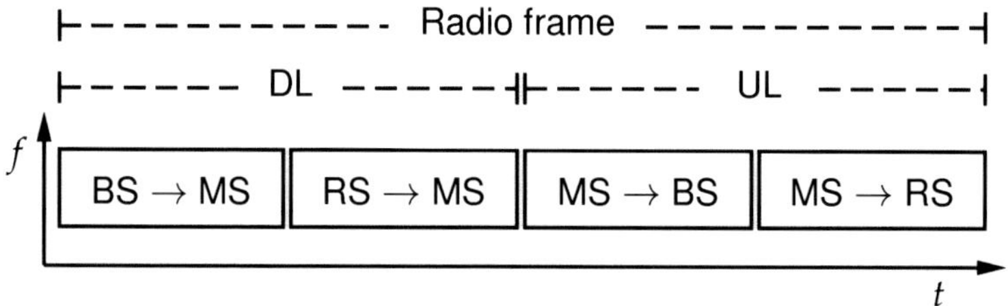

Figure 2.7: Relay Stations in a TDD System

required between DL and UL sub-frames as well as between phases in a sub-frame dedicated to BS and RS communication, respectively, to avoid inter-symbol interference resulting from radio-signal propagation delay.

2.2.2.2 Frequency Division Multiplex

Most cellular mobile radio systems operate on radio channels either assigned for UL or DL. Under FDM BS and RSs transmit alternating in TDM mode of operation. Both, DL and UL frequency channels carry radio frames divided into BS-controlled and a RS controlled phases.

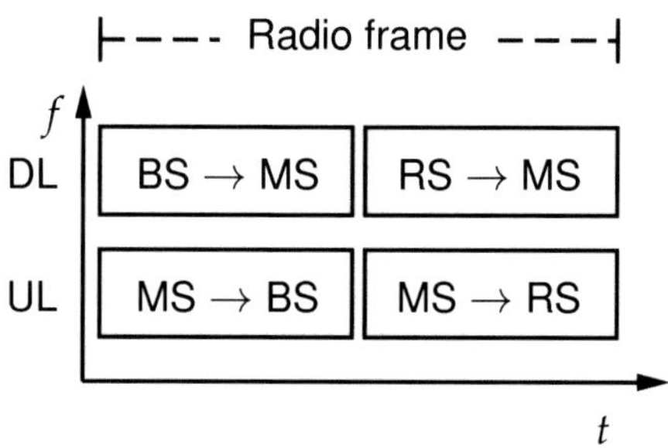

Figure 2.8: Relay Stations in an FDM system

Figure 2.8 shows how RSs are integrated in the radio frame. The MSs may operate in full-duplex mode, transmitting and receiving on different radio channels at the same time. Since this is costly in real world systems MS operate in half-duplex mode requiring a MS either to receive on DL or transmit on UL phases of the radio subframe.

2.2.3 Relay Placement in a Cell

Fixed relays appear to be beneficial to achieve maximum cell capacity in mobile broadband wireless radio networks (Sambale & Walke, 2012a). It turns out that three RSs per cell sector are necessary and sufficient to achieve the optimum capacity gain in the evaluation scenarios specified by International Telecommunication Union (ITU) (ITU-R, 2008a). More than three RSs increase cell capacity marginally but increase cost. Following ITU guidelines (ITU-R, 2008a) for the remaining of this work a cell is a 120 ° sector served by a BS site, see Fig. 2.9. Following (Sambale, 2013; Sambale & Walke, 2012a, 2012b) in each cell three RSs are placed, one in antenna boresight at a distance D_2 of 95 % of the cell radius and two RSs are placed left and right with an angle of $\alpha = 17\,^{\circ}$ to antenna boresight at a distance D_1 of 70.5 % of the cell radius R, see Fig. 2.9.

This RS placement is assumed for most of the traffic performance evaluation studies presented in Chapter 6. It is also shown there that deviations from this way of relay placement do not change the performance results much. In real world systems, and in the simulator presented in the following, radio coverage of BSs serving adjacent cells overlap at sector cell borders, shown by dotted lines, see (Sambale & Walke, 2012a) and overlap with sector cells of adjacent sites.

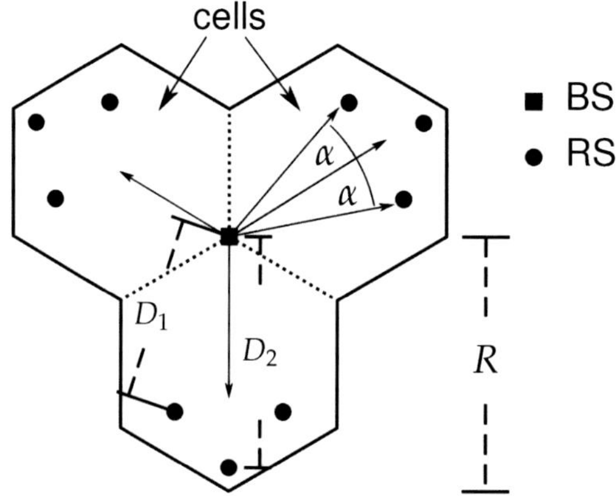

Figure 2.9: Relay placement

2.2.4 Related Work

Fixed relays to provide enhanced coverage in mobile radio networks have been proposed about 39 years ago (Walke & Briechle, 1985) and became a research subject around the year 2000 where cellular mobile broadband systems were recognized to need infrastructure based architecture support to achieve capacity expectations for the coming decades (Esseling, Vandra, & Walke, 2000; Lin & Hsu, 2000; Pabst et al., 2004). Since then, there have been numerous investigation and publications that propose RSs either as enhancements for existing networks or as an architecture element for cellular networks. RSs play an important role in 4G mobile radio networks since standardized in 3GPP long term evolution (LTE) Release 10 and in IEEE 802.16m.

In (Pabst et al., 2004), it is shown how fixed relays can extend the range and combat shadowing in urban areas as well as improve the system capacity. Furthermore it shows how RSs reduce infrastructure deployment costs. (Esseling et al., 2005) shows a delay and throughput analysis of relay enhanced wireless broadband systems. The investigation outlines IEEE 802.11e, 802.15.3, 802.16a and HIPERLAN/2 DL performance in a Manhattan like scenario. (Nourizadeh, Nourizadeh, & Tafazolli, 2006) shows that RSs improve the capacity of an Universal Mobile Telecommunication System (UMTS) frequency division duplex (FDD) system by 26% for VoIP traffic load by placing six RSs inside the cell coverage area.

CHAPTER 3

IEEE 802.16 System

Contents

The modular software platform introduced in this work serves to represent the IEEE 802.16 based system. To ease understanding of functioning of such systems this chapter provides an introduction into the specifications of this IEEE standard.

3.1 The ISO/OSI Reference Model

The International Standardization Organization (ISO)/Open System Interconnection (OSI) recommendation provides a common basis for the coordination of standards development for the purpose of systems interconnection (ISO, 1994). Both, the ISO/OSI reference model and the IEEE 802.16 protocol are based on a layered architecture. Each layer offers a service to the next higher layer through the service access point (SAP). A communication between two communication partners is only performed between so-called entities on the same protocol layer with the help of the service of the next lower layer. Layers offer services to higher layer entities through SAPs, see Fig. 3.1 (Zimmermann, 1980).

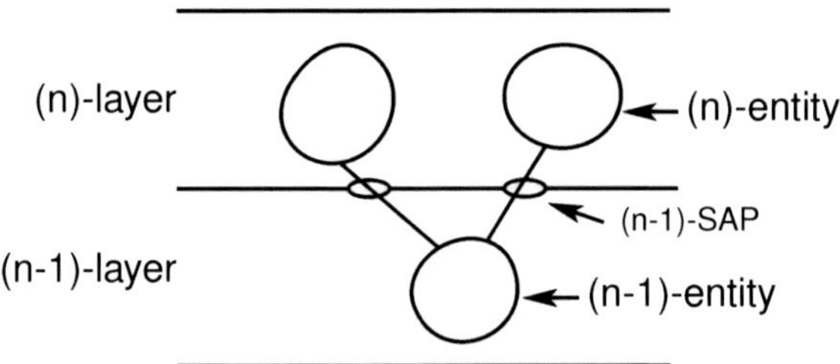

Figure 3.1: Layers, Entities and SAPs

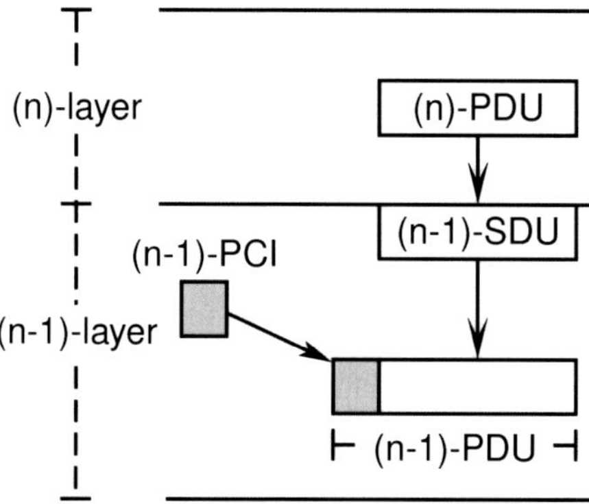

Figure 3.2: Data units in adjacent layers

Protocol layer exchange data between entities of the involved systems in form of protocol data units (PDUs) that contain layer specific protocol control information (PCI) and the service data unit (SDU) of the next higher layer. PCI is control information that exchanged between two n-entities using (n-1)-connections, see Fig. 3.2. The (n-1)-layer encapsulates the (n-1)-SDU that is virtually the (n)-PDU, appends the (n-1)-PCI and thereby creates the (n-1)-PDU. The peer (n-1)-entity separates the (n-1)-PCI and delivers the (n-1)-SDU to the next higher layer. For layer internal control functions like connection establishment, say for a logical channel, or synchronization the (n-1)-entity creates a (n-1)-PDU without a (n-1)-SDU. In such a case, the PDU contains PCI only.

The ISO/OSI reference model defines seven layers, see Fig. 3.3.

The *application layer* serves the application process as SAP for the OSI. It offers means to initiate, maintain and terminate connections between application processes. It defines the protocol by which peer application processes communicate via the OSI.

The *presentation layer* provides the set of services that enables the application layer to interpret the meaning of the exchanged

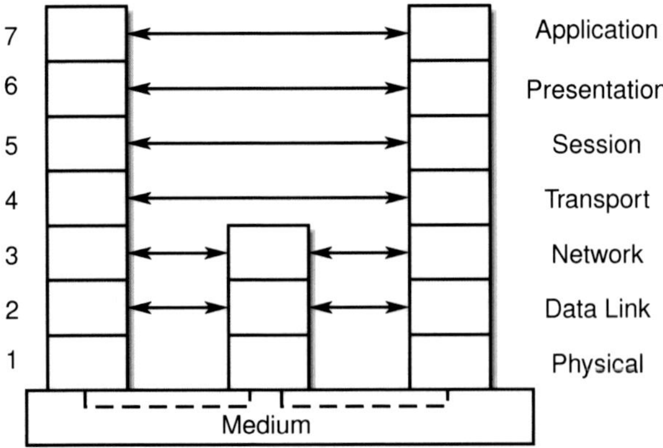

Figure 3.3: ISO/OSI Reference Model Architecture

data. For this the presentation layer defines a syntax that is common for both entities and that enables the layer to convert the data between different formats.

The purpose of the *session layer* is to administer a session service that is independent from the underlying transport connection. The session layer establishes one or more subsequent transport layer connections and maps the session onto the transport connection.

The *transport layer* provides a transparent transport service for the session layer without the concern of how the network establishes the connection to the peer entity. Typical functions of the transport layer are flow control, congestion control and segmentation.

The *network layer* establishes a point-to-point (PtP) connection via the data link or a sequence of data links. The layer protocol implements routing functions for providing a path to the destination node.

The *data link layer* operates on the bit-stream of the physical layer and encodes or decodes data link layer frames. It also guarantees reliable transmission of the data link PDU to the peer entity. To eliminate transmission errors the data link layer uses error detection and correction codes and applies an automatic repeat request (ARQ) protocol for retransmission of erroneously received link layer PDUs.

The *physical layer* establishes a connection to the a peer entity for bit-wise transmission over a given medium. In wired networks, the physical layer defines electronic parameters for interfaces at connectors, synchronization patterns, power levels, signal coding and modulation.

3.2 Scope of standard IEEE 802.16

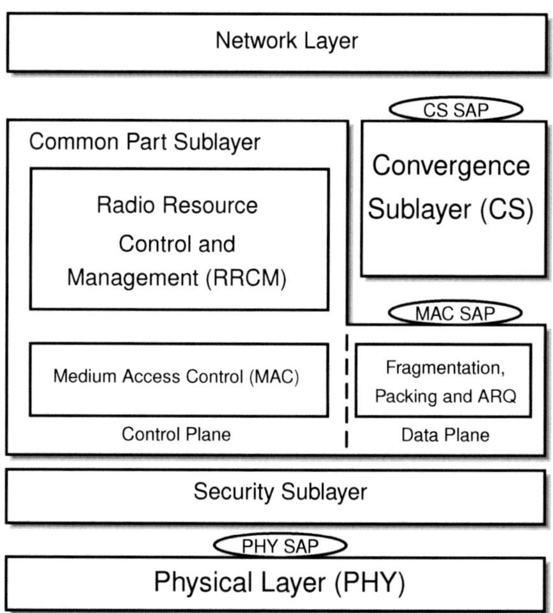

Figure 3.4: IEEE 802.16 Protocol Structure (Srinivasan & Hamiti, 2010)

The IEEE 802.16 standard specifies the protocol stack below the network layer that is subdivided into sublayers. The highest sublayer comprises two parts, see Fig. 3.4. The convergence sublayer (CS) provides a SAP for network layer services and acts as an adaption layer between network layer protocols and the common part sublayer (CPS). The standard defines the IP-CS for the Internet protocol (IP) and the ATM-CS for networks based on asynchronous transfer mode (ATM). The purpose of the CS is to classify network layer packets according to the payload, namely the packet content, and therefore depends on the type of network layer.

Following the ISO/OSI reference model the IEEE 802.16 standard classifies the functions of its protocol stack into three categories, namely data plane, control plane and management plane. The data plane comprises protocol functions and services of payload data processing such as header compression segmentation of payload data which means dividing a payload SDU into smaller segments. The control plane comprises protocol functions that involve radio resource configuration, coordination signaling and management. The management plane comprises protocol functions and services for external management and configuration of the network. For this work, the management plane is important for MS network entry but not shown in Fig. 3.4.

The CPS is common to all CSs and independent of both, the type of the CS and the physical (PHY) layer. The CPS comprises radio resource control and management (RRCM) functions and medium access control (MAC) functions. Figure 3.4 shows the most important parts of the CPS and how the CPS and the CS integrate into the protocol stack.

The *radio resource management* function of the RRCM adjusts radio network parameters based on traffic load and performs load balancing, admission control and interference control. The *mobility management* function comprises both, intra-system handover between BSs and RSs and inter-system handover to other radio access technologies.

3.2.1 Network Entry and Connection Management

The protocol distinguishes the BS as the service provider of the radio network from the MS that acts as the service user. To gain access to the network the MS request the network entry at the BS as described in the following.

The MS chooses the BS with the strongest received signal strength among a set of BSs. For BS search, the MS uses the preamble of the *radio super-frame* transmitted on a broadcast control channel to size the signal strength. After the ranging process has been completed, the MS negotiates basic capabilities with the BS which is followed by the optional authentication, authorization and key exchange. As soon a MS finds a network, the MS establishes a connection to the network as shown in Fig. 3.5.

Network entry includes ranging and UL synchronization, basic capability negotiation, authentication, authorization, key exchange, registration at the BS and service flow establishment as usual with wireless and mobile systems.

Ranging RNG is the process to acquire a timing offset, frequency offset and power adjustment that the transmission of the MS align with the BS receive frame. The BS offers initial ranging intervals in the UL control channel and indicates them in the UL-medium access pointers (MAP) messages. The initial ranging interval indicated in the UL-MAP contains one or two transmission opportunities that allow the MS to send ranging requests (RNG-REQs) to the BS. The MS selects a ranging slot of the initial ranging interval and sends a RNG-REQ to the BS. The BS sends a ranging response (RNG-RSP) to the MS that indicates the successful reception of the RNG-REQ. The response contains

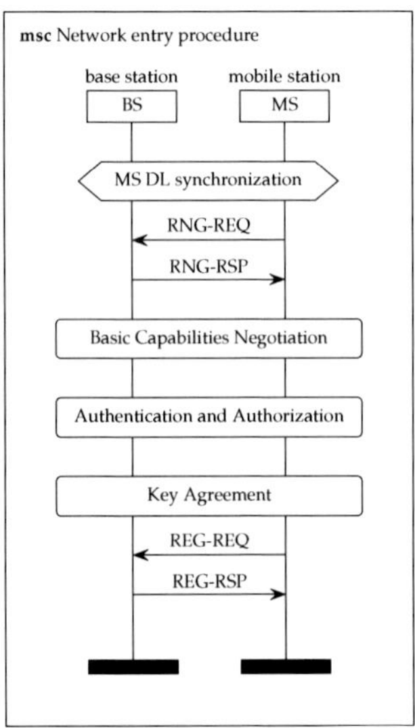

Figure 3.5: Network entry procedure

further time synchronization adjustment information if needed. After the ranging process has been successfully completed, the MS awaits an UL bandwidth allocation, see Section 3.4.2, for basic capabilities negotiation and performs authorization procedures.

After the MS has been authorized to connect to the network, the MS requests the network entry with the register request (REG-REQ). The BS grants the request with the register response (REG-RSP). After registration, the MS establishes IP connectivity via dynamic host configuration protocol (DHCP).

3.3 Medium Access Control

The MAC of the IEEE 802.16m system is the data link layer (DLL) equivalent of the ISO/OSI reference model. The MAC entity in the BS controls the access to the radio resource for the whole cell.

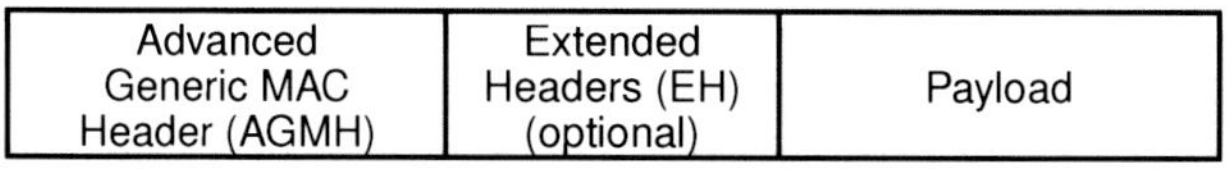

Advanced Generic MAC Header (AGMH)	Extended Headers (EH) (optional)	Payload

Figure 3.6: MAC Frame Format

Figure 3.6 shows the basic MAC frame of the MAC sublayer protocol. Each frame carries the advanced generic MAC header (AGMH) and optionally may contain one or more extended headers (EHs) followed by the payload. Table 3.1 shows the contents of the AGMH.

Table 3.1: Advanced Generic MAC Header

Syntax	Size [bit]	Description
FID	4	flow identifier
EH	1	indicates whether EH is present in MAC frame
length	11	length of the MAC frame including header in bytes

The service flow identifier (SFID) or flow identifier (FID) is unique for a MS and identifies the service flow the MAC frame belongs to. The BS defines the FID on connection establishment, see Section 3.3.3.

Table 3.2 shows the EH types that are available in the protocol. Fragmentation and packing extended header offer the option to indicate either fragments of a MAC SDU or packed fragments of multiple MAC SDUs, see Section 3.3.5 for further details on ARQ operation. The MAC control EH indicate control messages that require an acknowledgment. Multiplexing EH

Table 3.2: Extended Header Types

ID	Name
0	MAC SDU fragmentation
1	MAC SDU packing
2	MAC control
3	multiplexing
4	MAC control ACK
5	piggybacked bandwidth request
6	rearrangement fragmentation and packing
7	ARQ feedback polling
8	advanced relay forwarding

indicates the presence of multiple connection payloads in the MAC PDU that share the same security association. The MAC control acknowledgment (ACK) indicates an acknowledgment to a MAC control message. Piggybacked bandwidth request EH indicates the presence of a bandwidth request in a user data MAC PDU, see Section 3.3.6. Rearrangement fragmentation and packing EH indicate the presence of an ARQ sub-block that is constructed from SDUs or SDU fragments or both that are packed. It is used for ARQ enabled transport connections frames only that require a re-transmission of a previously lost MAC PDU with packing extended header (PEH). The ARQ feedback polling EH shall be used by an ARQ transmitter to poll ARQ feedback. The advanced relay forwarding EH may be used when data traffic between BS and RS is encapsulated into MAC PDU. For further details about the EH types, see (IEEE, 2012).

3.3.1 Connection Types

In the Internet represented by the network layer in Fig. 3.4 a connection-less service is used. In a wireless system like IEEE 802.16m serving to bridge the gap between a MS and fixed network elements a connection oriented network service is required to be able to differentiate packets transmitted on a radio link established beforehand. Virtual connections are used that are identified by their connection identifier (CID) in the wireless network. A connection in terms of the IEEE 802.16m system is a mapping between a BS and one or more MSs. A unicast connection maps a BS to a single MS. A mapping to multiple MSs is called either multicast or broadcast connection depending on whether multiple or all MSs is addressed by the connection, respectively.

A number of virtual connections are to differ each characterized by its distinct range of values of the CID.

Initial Ranging Connection

Initial ranging is performed as part of the association process to the BS by the MS before the MS establishes a connection to the BS. Since a MS does not know any unused CIDs the static *initial ranging* CID is reserved for MSs that intend to enter the cell.

Basic Connection

The BS assigns a single basic connection for each MS. The basic connection carries time critical MAC management messages for both, UL and DL.

Primary Management Connection

The *primary management* connection carries delay-tolerant MAC management messages like service flow related management messages, privacy key management messages and multicast group assignment management messages.

Secondary Management Connection

The *secondary management connection* carries management messages of higher layers like simple network management protocol (SNMP) or DHCP. The BS assigns a secondary management connection if the MS identifies itself as *managed MS*.

Transport Connection

Transport connections carry payload frames for either UL or DL. The BS may assign multiple transport connections to a MS. Since transport connections are unidirectional, the BS usually assigns pairs of transport connections. Unidirectional self-contained transport connections are possible.

In addition to the connection types shown above, the standard defines several connection types that are less relevant for this work and are not presented here.

3.3.2 Addressing

Each mobile node that contains a IEEE 802.16m entity, also known as mobile station (MS), has a unique MAC address of 48 bit. MSs use the MAC address to identify themselves at the BS. During association of a MS to the network the BS establishes a connection to the MS that is identified by a CID of 16 bit. Both, BS and MS use the CID as a short reference in its layer-2 PDUs for communication following the initial ranging end network entry.

3.3.3 Service Flows

When a MS intends to send user data, the BS provides a service flow that is identified by a SFID. The BS binds the service flow to a transport connection, identified by a transport CID and the requested QoS parameters. Multiple service flows may be multiplexed to a single wireless transport connection. The CS keeps track of the established service flows and maps a packet to a service flow with the help of classification rules. These rules are based on the content of the packet and may take target network address, target or source transport protocol address into account.

Service flows support QoS on per-connection basis. As a result service flows with different QoS parameters require different connections. The following list shows some of the QoS parameters for service flows.

Maximum sustained data rate The parameter defines the peak information data rate that the service flow should provide. However, it doesn't specify a guarantee, that the data rate is available. This parameter should match the average data rate of the traffic load over time.

Maximum traffic burst The *maximum traffic burst* parameter define the maximum short-term burst size of the traffic load. It specifies the maximum continuous burst size the system should accommodate for the service flow.

Minimum reserved data rate The *minimum reserved data rate* parameter specifies the data rate the system must be able to provide at any time for the service flow. This parameter is taken as a guaranteed minimum data rate.

Maximum latency The *maximum latency* parameter specifies the upper bound of the interval a MAC SDU takes to travel from the CS-SAP of a sending station to the CS-SAP of a receiving station. The system must provide the maximum latency guarantee for all packets of the service flow.

UL grant service type The parameter specifies the *UL grant service type* for the service flow, see Section 3.3.4. The parameter is only available for UL service flows.

3.3.4 UL Grant Services

UL request and grant scheduling provides each MS with bandwidth for UL transmissions. Granting UL resources is performed by the BS and is indicated in the UL-MAP as transmission opportunities. UL resources are either provided for UL user PDUs or UL control PDUs, namely bandwidth requests. Along with the grant service type and the QoS parameters of a service flow the BS is able to allocate UL radio resources units at the appropriate time.

With a wide range of traffic services in mind, the IEEE 802.16m system provides a set of uplink resource grant functions and services.

3.3.4.1 Unsolicited Grant Service

The unsolicited grant service (UGS) provides fixed-rate resource grants on a periodic real-time basis by means of a TDM channel that eliminates the explicit signal overhead required for resource requests. The BS allocates UL transmission opportunities based on the minimum reserved traffic rate of the service flow. The size of the reservation counted in resource units shall be sufficient to contain the fixed-length MAC frames that is associated with the service flow. The unsolicited grant service is designated for constant bit-rate services like VoIP without silence suppression.

3.3.4.2 Real-time Polling Service

The real-time polling service (rtPS) supports UL service flows of real-time applications with variable size data packets. The service provides periodic unicast request opportunities which meet the flow's latency and data rate requirements. The service requires more signal overhead than the UGS service but in turn supports variable sized packets. A typical payload of the rtPS is video streaming with a variable bit rate .

3.3.4.3 Extended Real-time Polling Service

The extended real-time polling service (ertPS) combines the reduced overhead requirements of UGS and the flexibility of rtPS. The BS provides periodic unicast grants without a dedicated bandwidth request. Unlike with UGS the size of the grant of ertPS is dynamic. The BS allocates UL transmission opportunities that can either be used for payload transmissions or bandwidth change request that are indicated by the extended piggyback request field of the grant management subheader (GMSH) or by the bandwidth request (BR) field of the MAC signaling header.

3.3.4.4 Non Real-time Polling Service

The non-real-time polling service (nrPS) provides unicast polls on a regular basis. As a result the UL service flow is able to request bandwidth even during network congestion. Non real-time service flow allows for delay-tolerant data streams that require a minimum data rate.

3.3.4.5 Best Effort Service

Services that do not require specific latency or data rate requirements are scheduled by the best effort grant service. The MS is allowed to use contention access opportunities for bandwidth requests and unicast access opportunities for data transmissions. The BS does not poll MSs for bandwidth requests.

3.3.5 Automatic Repeat Request Protocol

The logical link control (LLC) protocol provided in the data plane supports ARQ operation on a LLC-connection basis using blocks to transmit PDUs. ARQ type and parameters are negotiated during connection setup. The ARQ protocol provides in-order delivery according to the ARQ block sequence number (SN). Multiple MAC SDUs may be packed into a MAC PDU if the SDU size is smaller than the maximum ARQ block size.

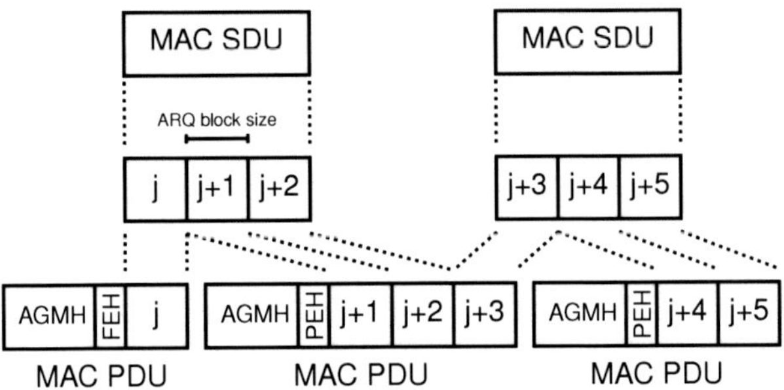

Figure 3.7: ARQ fragmentation and packing

Figure 3.7 shows fragmentation of MAC SDUs and packing of ARQ blocks into MAC PDUs. SDUs are cut into fragments that do not exceed the maximum ARQ block size. The PDU accumulates multiple fragments in a single PDU. MAC PDU that contain multiple segments of a MAC SDU are indicated by the fragmentation extended header (FEH). Table 3.3 shows the contents of the FEH.

In Table 3.4, fragmentation control (FC) specify whether multiple ARQ blocks are packed into a MAC PDU or if the ARQ

Table 3.3: Fragmentation Extended Header

Syntax	Size [bit]	Description
Type	4	extended header type: FEH = 0
FC	2	fragmentation control bits
SN	10	ARQ block sequence number for ARQ supported connections; MAC PDU sequence number counter otherwise

blocks come from a single SDU only.

If the MAC PDU consists of fragments of multiple SDUs, the fragments are indicated by a PEH that specifies the number of fragments and their length according to Table 3.4.

Table 3.4: Packing Extended Header

Syntax	Size [bit]	Description
Type	4	extended header type: PEH = 1
FC	2	fragmentation control bits
SN	10	ARQ block sequence number for ARQ supported connections; MAC PDU sequence number counter otherwise
length	11	Length of the SDU or SDU fragment.
end	1	0 = another *length* and *end* field is following; 1 = no more *length* and *end* field is following
padding		byte alignment

3.3.6 Bandwidth Request

Depending on the service grant, bandwidth requests in the IEEE 802.16m system are either granted by radio resource control on a data rate or a data volume basis. For polling services the MS requests UL opportunities for a given amount of data as soon as it is polled by the BS.

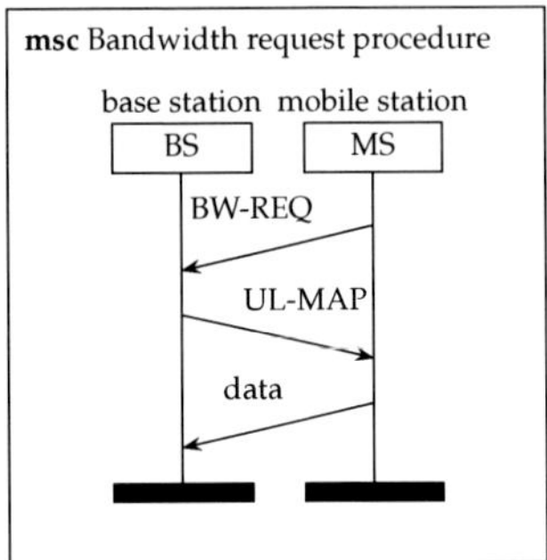

Figure 3.8: Bandwidth request procedure

As Fig. 3.8 shows, the MS sends a BW-REQ to the BS that contains the station identifier (STID) of the requesting MS, the FID for the requested connection and the requested data volume to be transmitted. According to the QoS context of the flow, the BS acknowledges the bandwidth request with UL opportunities for the MS in the UL-MAP. Upon the grant of UL opportunities the MS sends the MAC PDU to the BS.

UL service grants that are based on data rate are requested at the beginning of the service session. With ertPS, the UL service grant can be adapted, if the traffic source requires increased data rates.

3.4 Physical Layer

The IEEE 802.16m standard defines several PHY profiles for different use cases. A single carrier PHY profile specified for frequencies from 10-66 GHz is a designated LoS profile for short range links, as usual with radio relay systems. The orthogonal frequency division multiplex (OFDM) physical layer is specified for frequencies below 10 GHz for operation in use cases where multi-path signal propagation and spacial shadowing is present, like in the 802.16m metropolitan area network application for service to fixed subscriber stations.

The orthogonal frequency division multiple access (OFDMA) profile is specified for use cases where multi-path propagation and shadowing is dominating, namely the mobile Worldwide Interoperability for Microwave Access (WiMAX) extension and the 802.16m mobile broadband system meeting the IMT-A requirements. In the following we only focus on the OFDMA physical layer.

3.4.1 Duplex Modes

The OFDMA profile supports both duplex schemes, time division duplex (TDD) and frequency division duplex (FDD). The FDD mode is foreseen to be used in paired licensed frequency bands. TDD allows to dynamically adapt the DL/UL ratio of radio resources assigned in a radio frame according to the traffic load demands. It operates on a single frequency band only.

3.4.1.1 Frequency Division Duplex

Since the bandwidth assigned to a radio channel is fixed, the DL/UL bandwidth ratio is fixed and can not be modified during operation. In FDD operation stations can transmit on one radio channel and receive on the other simultaneously since the channels are sufficiently spaced in the frequency domain. Since FDD duplex-filter are expensive mobile stations mostly operate in half-duplex mode for economy reasons, so that MSs do not transmit and receive on different frequency channels at the same time.

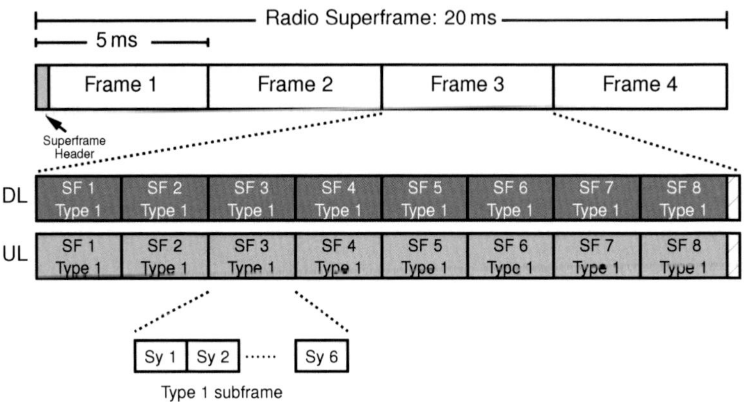

Figure 3.9: IEEE 802.16 FDD Radio Frame

Figure 3.9 shows the FDD OFDMA radio frame for 10 MHz. The superframe of 20 ms is divided into four frames of 5 ms and transmitted periodically in time on the radio channel. DL and UL frames are further divided into eight *Type-1* sub-frames. Table 3.5 show that each Type-1 frame carries 6 OFDM symbols, as a result the 5 ms-radio frame sums up to 48 OFDM symbols.

Since the sampling frequency differs for different bandwidth settings, the number of OFDM symbols does not always sum up to a divisible number. The PHY layer meets the issue by aggregating different sub-frame types of different lengths in a frame. Table 3.5 shows the sub-frames available in the PHY layer. Each bandwidth setting requires an individual frame setup that configures different sub-frame types in the frame.

Table 3.5: OFDMA sub-frame types

Type	Symbols
Type-1	6
Type-2	7
Type-3	5
Type-4	9

3.4.1.2 Time Division Duplex

TDD allows for a flexible DL/UL ratio adaption. Unlike the FDD mode, the BSs of the cellular system need to be synchronized when using TDD to prevent interference from neighbor cells. If two cells are not synchronized, UL transmissions of a cell-edge user of one cell would heavily interfere the DL transmission to the cell-edge user of the other cell. Synchronization of cells is usually performed by Global Positioning System (GPS) time signals.

Figure 3.10 shows the basic structure of the TDD frame for a channel bandwidth of 10 MHz. The 20 ms superframe is time-periodically transmitted on the radio channel and is divided into equally sized 5 ms frames. With the channel bandwidth of 5 MHz, 10 MHz and 20 MHz a frame comprises eight subframes. With the channel bandwidth of 8.75 MHz the frame comprises seven subframes.

Table 3.6 summarizes the parameters of the OFDMA PHY layer for the bandwidth configuration of 5, 7, 8.75, 10 and 20 MHz.

3.4.2 Control Signaling

The BS carries out radio resource allocations for both, DL and UL channels. The BS informs the MSs about the resource al-

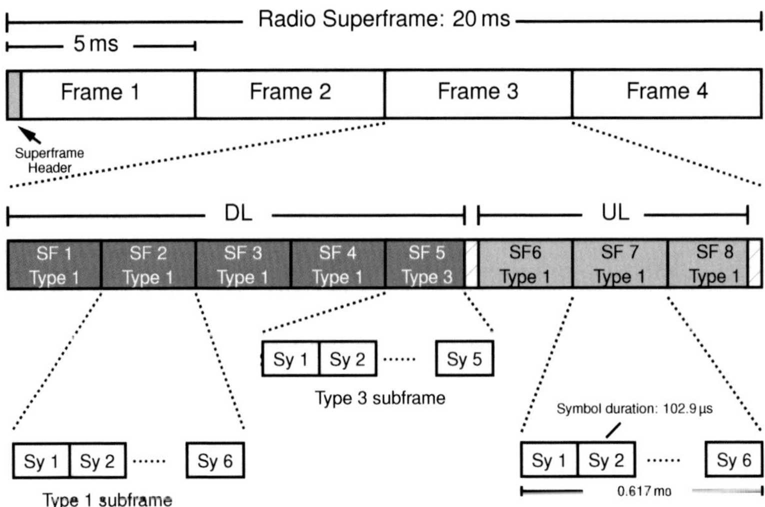

Figure 3.10: IEEE 802.16 TDD Radio Frame

location by Assignment-MAP in the MAP region in the DL sub-frames.

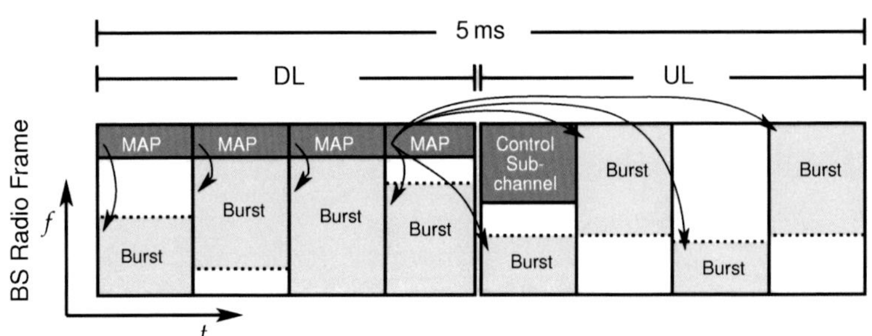

Figure 3.11: OFDMA TDD-Radio Frame with Control Channels

Figure 3.11 shows that the MAP control channels for DL direction are located in dedicated subchannels on each sub-frame.

Table 3.6: OFDMA radio channel parameters

Channel bandwidth [MHz]		5	7	8.75	10	20
Sampling frequency F_S [MHz]		5.6	8	10	11.2	22.4
FFT size N_{FFT}		512	1024	1024	1024	2048
Subcarrier spacing Δf [kHz]		10.94	7.81	9.77	10.94	10.94
Useful symbol time T_b [µs]		91.4	128	102.4	91.4	91.4
OFDMA symbol time T_S [µs]		102.857	144	115.2	102.857	102.857
FDD	Number of OFDMA symbols per 5 ms	48	34	43	48	48
	Idle time [µs]	62.857	104	46.4	62.857	62.857
TDD	Number of OFDMA symbols per 5 ms	47	33	42	47	47
	TTG + RTG [µs]	165.714	248	161.6	165.714	165.714

Among others, the MAP contains hybrid automatic repeat request (HARQ) feedback information elements (IEs), power control IEs and assignment IEs. The assignment IEs indicate the radio resource allocation for both DL and UL. Although Fig. 3.11 shows resource assignment IEs to indicate the resource allocation of the current frame, DL and UL assignment IEs are valid for the following frame instead.

UL control signaling by MSs has to be transmitted in UL transmission opportunities, assigned by the BS. For time critical messages like HARQ feedback, MSs transmit in the dedicated control sub-channel, shown in Fig. 3.11.

3.4.3 Physical Radio Resource Structure

The frequency band of a radio channel is divided into four or less frequency partitions. Each partition is furthermore divided in the time and frequency domain into contiguous and non-contiguous physical resource units (PRUs).

The PRU is the basic physical unit of the radio resource allocation. It comprises P_{SC}=18 adjacent subcarriers used to transmit

N_{sym} consecutive OFDMA symbols. N_{sym} is 6, 7 and 5 symbols for type-1, type-2 and type-3 subframes respectively, see Table 3.5. The logical resource unit (LRU) is the basic logical unit for the resource allocation and has the same size as the PRU.

Subcarrier permutation allows for allocation of subcarriers that are distributed over the entire frequency partition. Resource units that comprise distributed subcarriers from all over the frequency partition are called distributed logical resource unit (DLRU). DLRUs take advantage of frequency diversity gain. On the contrary, a contiguous logical resource unit (CLRU) comprises subcarriers that are aligned adjacently in a given frequency partition. CLRUs are known to achieve frequency selective scheduling gain (Einhaus, 2009).

Resource allocations performed by the radio resource scheduler are applied to LRUs. Mapping LRUs to PRUs is not the task of the radio resource scheduler but is fixed for the used PHY profile, represented by the type of subframe.

3.4.4 Hybrid ARQ

In modern mobile radio networks, the use of a HARQ is mandatory. It improves error correction by combining re-transmission of erroneously received data blocks with facilities of the channel coder. With *chase combining*, the re-transmitted data block is combined with the previous transmitted block. The combination of the two blocks adds information that may be sufficient to correctly decode the transport block.

With *incremental-redundancy-HARQ* instead of re-transmitting the same block another block is formed at the transmitter that carries other redundant information and both received blocks are combined by the channel coder.

3.4.4.1 Chase Combining

Two consecutive received transmissions of the same data frame are combined to increase the mutual information (MI). According to (Brueninghaus et al., 2005) Equation (3.1) gives the mutual information I_I after q transmissions. I_m is the mutual information function of modulation m. γ_{nj} is the SINR of the n-th symbol at j-th transmission. N denotes the number of all symbols contained in a data frame.

$$I_I = \frac{1}{N} \sum_{n=1}^{N} I_m \left(\sum_{j=1}^{q} \gamma_{nj} \right) \tag{3.1}$$

Chase combining requires re-transmissions of the same data frame with the same modulation and coding schemes (MCS).

3.4.4.2 Incremental Redundancy

A well known approach to increase the received MI of a transport block is to provide additional redundancy by retransmissions. For erroneous frames, the sender transmits frames with different redundancy bits to the receiver. From the combination of frames with different redundancy bits the decoder more safely may decode the received packet. Equation (3.2) gives the MI per bit after transmission of frame number $n+1$, see (Srinivasan, 2009).

$$I_{n+1} = \frac{N_{\text{pre}} \cdot I_n + N_{\text{NR}} \cdot \overline{I_b} + N_{\text{R}} \cdot f_1 \left(f_1^{-1}(I_n) + f_1^{-1}(\overline{I_b}) \right)}{N_{\text{pre}} + N_{\text{NR}} + N_{\text{R}}} \tag{3.2}$$

I_{n+1} is the mean mutual information after $n+1$ transmissions. N_{pre} is the number of bits that are not re-transmitted in the $n+i$-the transmission but have been transmitted before. I_n

is the mean mutual information of the bits up to the n-th re-transmission. N_{NR} is the number of newly transmitted bits. N_R is the number of bits that have been transmitted in previous transmissions and re-transmitted in the $n+i$-th transmission. $\overline{I_b}$ is the average mutual information per bit of the $n+1$-th transmission. Incremental redundancy allows to transmit the same frame multiple times with a different modulation and code rate.

With the IEEE 802.16 standard use of HARQ is mandatory for unicast data traffic in both directions, DL and UL. The HARQ operates as a send-and-wait ARQ protocol on multiple parallel channels.

3.4.4.3 DL Hybrid ARQ Procedure

HARQ in DL direction operates adaptive and asynchronously, which means a re-transmitted HARQ block may be allocated on a resource unit different from that used for the initial transmission. HARQ blocks are identified by the HARQ channel identifier (ACID).

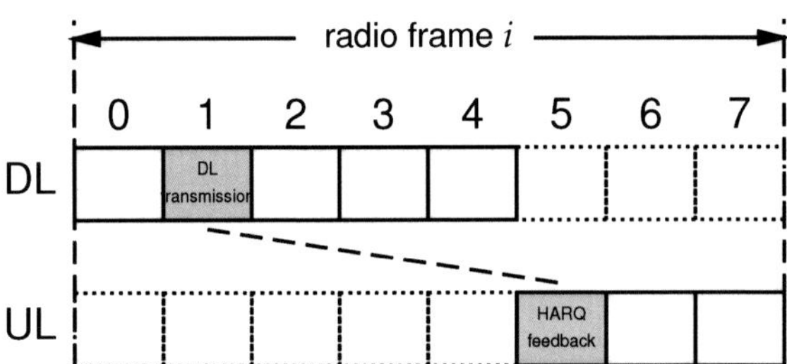

Figure 3.12: HARQ DL feedback timing for TDD

Figure 3.12 shows the timing of a successful HARQ transmission. The DL-MAP indicates a transmission block in sub-frame 1. If the MS is successful to decode this HARQ block it sends an ACK to the BS in sub-frame 5. Otherwise the MS sends a non-acknowledgment (NACK) to the BS in this subframe. The BS either re-transmits the block or sends a block with additional information in the next radio frame.

3.4.4.4 UL Hybrid ARQ Procedure

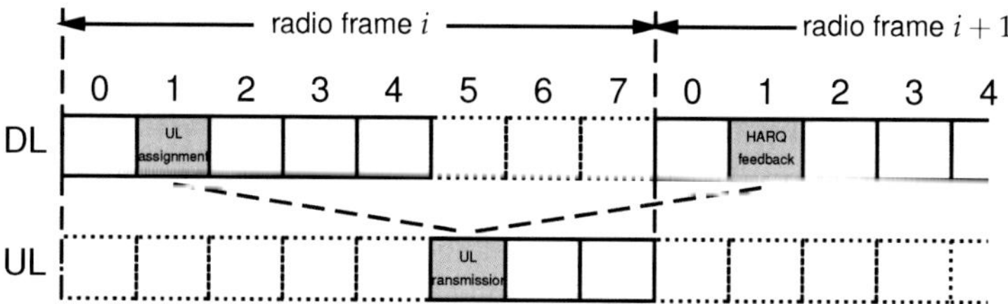

Figure 3.13: HARQ UL feedback timing for TDD

Figure 3.13 shows hybrid-ARQ operation in UL. After the BS has assigned UL opportunity for the MS in the UL access zone in subframe 5 it indicates the assignment in the UL MAP in one of the DL subframes, here subframe 1. The MS transmits the burst according to the assignment. In the next radio frame, the BS in subframe 1 sends the HARQ feedback message in the DL access zone on the MAP control channel. If the transmission failed, the BS sends a NACK and assigns a new UL transmission opportunity to be used by the MS.

3.5 Network Reference Model

The WiMAX Forum defined a network model (WiMAX Forum, 2009) that has been adopted by the IEEE as the 802.16 network model. The WiMAX network model is based on the design principles shown below.

IP service optimized access service network (ASN)

It is expected that more and more services in 4G systems are based on the IP protocol. Real-time voice, multicast and broadcast of multimedia and conventional web services will dominate the payload of future wireless radio network systems.

IP interconnected ASNs

The network should be flexible in network design and realization. Redundancy and scalability are provided by IP-based interconnectivity while allowing the use of cost-efficient high bandwidth wireline or wireless backhaul components. Functional architectures and topology are based on Internet design principles.

Logical separation of ASN, CSN

Separation of network components allows to share the ASN among two or more connectivity service network (CSN) providers. Third party service providers should also be able to provide broadband IP services via two or more CSN operators.

Technology interoperability

Mobile broadband systems must accommodate heterogeneous technologies. Consistent interfaces based on IP protocol allow to ensure seamless access and connectivity. Furthermore, open interfaces and standards will permit evolution to future mobile broadband technologies.

Figure 3.14 shows the network elements, defined by the WiMAX Forum. The network access provider (NAP) offers the network access via the radio interface defined by standard IEEE 802.16. The network reference model allows the MS to use a foreign network service provider (NSP). The foreign NSP performs authentication, authorization and accounting (AAA) for

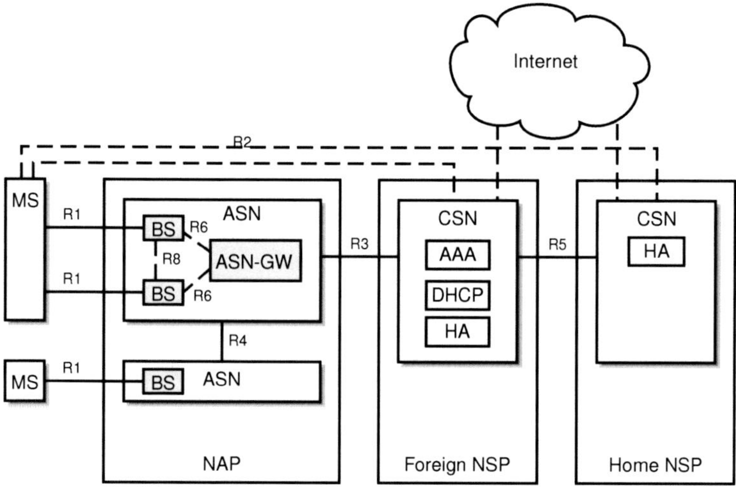

Figure 3.14: WiMAX Network Reference Model

MSs with the help of the home NSP home agent (HA). User data services are either provided by the home NSP or the foreign NSP if the MS associates to a foreign network.

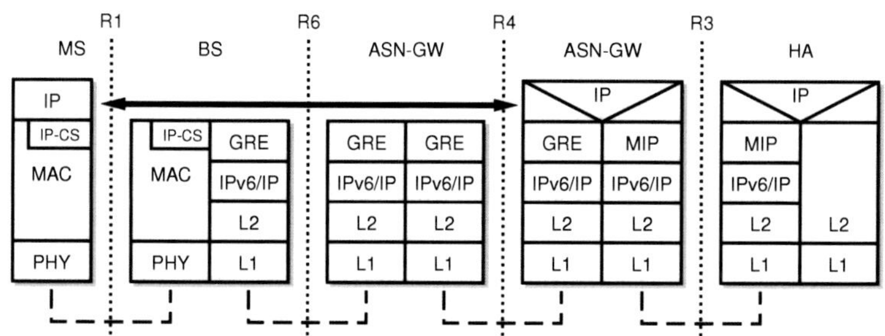

Figure 3.15: WiMAX User Data Protcocol Stack

Figure 3.15 shows the protocol stack of the IEEE 802.16 network model (Andrews, Ghosh, & Muhamed, 2007, Chap. 10).

The R1 interface connects the MS to the BS. Both have protocol stacks as shown in Fig. 3.4. The BS maps an 802.16 connection identified by a CID on the generic routing encapsulation protocol (GRE) that tunnels data at the R6 interface for both, down- and upstream traffic. There is a 1-to-1 correspondence between the 802.16 connections and the GRE keys that identify tunnel connections in the GRE protocol.

3.6 Multi-Hop Operation

In July 2005, the mobile multi-hop relay study group for the IEEE 802.16 standard started the attempt to develop a protocol enhancement to support decode-and-forward relays in the system. In 2007, the task group published the 801.16j draft that introduced relay stations in the system the first time. The draft was not accepted in the market and was superseded by the 802.16-2009 standard which includes relay stations as proposed in (Pabst et al., 2004).

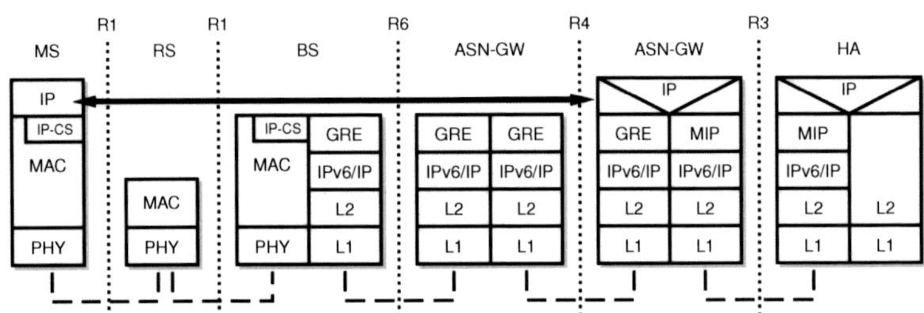

Figure 3.16: Protocol stack of the relay enhanced 802.16 system

As Fig. 3.16 shows the RS integrates in the system without a change to its protocol stack. From the MS point of view, the RS acts like a conventional BS. It grants or denies access to the network, allocates radio resources for UL and DL transmissions and keeps track of established connections. From the BS point of view, the RS acts like a MS with the accumulated traffic load of all of its client MSs. Therefore, it needs the respective radio resource allocation exceeding that of conventional MSs. Like the BS the RS needs reserved radio resources to communicate with its MSs. The RS of the IEEE 802.16 system does not route packets, like IP routers do. Instead the RS forwards UL frames to the BS and DL frames to the MS in layer 2, thereby operating as a bridge.

3.6.1 Relay enhanced MAC Frame

IEEE 802.16j task group based on (Pabst et al., 2004) developed a protocol extension for RSs that leaves the MS protocol unchanged. RSs integrate into the system seamlessly and MSs are not affected by relaying related system functions. The authors of (Tao, Li, Teo, & Zhang, 2007) with reference to the work of this author introduced a way to configure the radio frame to achieve system performance gain compared to the single-hop IEEE 802.16e system leaving the superframe structure unchanged.

The system allows for two different modes of RS operation. With the centralized resource scheduling, the BS allocates resources for both, transmissions from RS to MS and vice versa. For RS's subordinate RSs, resource allocations are also taken at the BS. Thereby the BS takes the scheduling load from the RSs to the BS. As a result the RS can be equipped with a lightweight processor.

Relaying in the IEEE 802.16 system is supported for both, TDD and FDD. The RS operates either in time-division-transmit and receive (TTR) mode, where relay and access link transmissions are multiplexed in time within a single frequency channel or in simultaneous transmit and receive (STR) mode, whereby the access and relay transmissions are performed simultaneously. The STR mode requires a strong separation of relay and access link antennas.

The IEEE 802.16 system supports non-transparent relays only, which means the RS is visible to MSs as fully functional BS. The BS assures that the RS can communicate to its subordinate stations without interference by reserving dedicated radio resource to the RS.

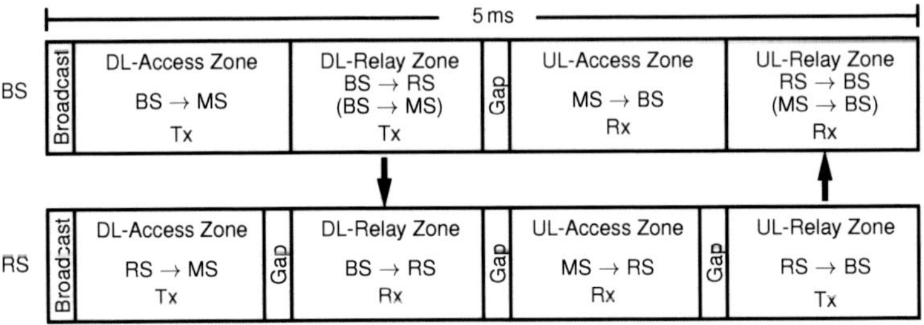

Figure 3.17: Relay enhanced Radio Frame

Figure 3.17 shows the relay enhanced MAC frame of the IEEE 802.16m amendment, according to (802.16m Task Group, 2009). The MAC frame is divided into the DL-Access zone, the DL-Relay zone, the UL-Access zone and the UL-Relay zone. The zones distinguish the transmission direction of the BS and the RS. In the DL and UL-Access zones the BS and the RS transmit or receive data to or from their associated MS respectively. In the relay zones the BS transmits or receives data frames to or from the RS only. During the relay zone the BS may also transmit or receive data frames from single-hop MSs.

From the MS point of view, non-transparent RSs act like BSs. MS can associate to non-transparent RSs and the non-transparent RS schedules the communication between the RS and its associated MSs. From the BS point of view, the RS acts like a regular MS.

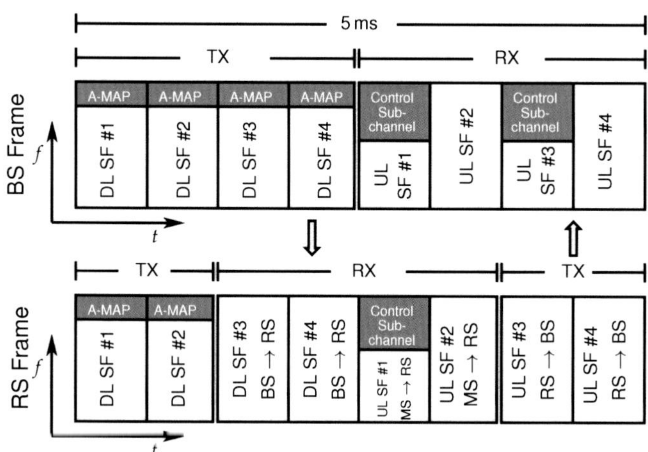

Figure 3.18: Example configuration with non-transparent RS

Figure 3.18 shows the relay enhanced radio frame for non-transparent RSs. Like the BS frame, the RS frame consists of sub-frame that are either dedicated DL or UL frames. Different from the BS frame, the RS receives data packets in the DL sub-frames that are part of the DL relay zone and sends packets in the UL sub-frames that are part of the UL relay zone. As a result the RS switches from transmit (TX) mode to receive (RX) and back to TX again.

3.6.2 Radio Resource Partitioning among Relay Stations

Balanced resource partitioning for RSs and BSs is essential for maximizing capacity. Unbalanced resource partitioning leads to early saturation of resources assigned to one hop and waste of resources on the other hop. Since the partitioning of radio resources can not be adapted for a single cell, the resource partition needs to be set for a cluster of closely connected cells. The system offers a set of possible resource partition configurations. Table 3.7 shows the number of sub-frames assigned for access and relay zones, respectively for DL and UL according to the frame configuration index. The configuration of the frame partitioning is identified by the frame configuration index (FCI).

Table 3.7: Frame configuration for RS operation

FCI	duplex mode	access zone		relay zone	
		DL	UL	DL	UL
0	TDD	3	1	3	1
1	TDD	3	2	2	1
2	TDD	2	2	2	2
4	FDD	4	4	4	4
17	TDD	3	1	3	1
18	TDD	2	1	3	2

RSs may reuse on the same UL and DL resources in the relay zone if the interference of neighbor RSs is negligibly small. Otherwise resources for RSs need to be separated either in time or in frequency.

3.6.3 Connection Management

Connections of MSs served by RSs are controlled by the BS. The BS establishes one or more tunnels to the RS that are identified by unique flow IDs. The BS may establish multiple tunnel connections to provide different QoS to different connections.

Figure 3.19 shows the binding of multiple connections that share the same QoS level to a single tunnel between the RS and the BS.

Multiple tunnels may be established between BS and RS after the network entry. Each tunnel is identified by a unique FID. Connections of a MS may be mapped on one or multiple tun-

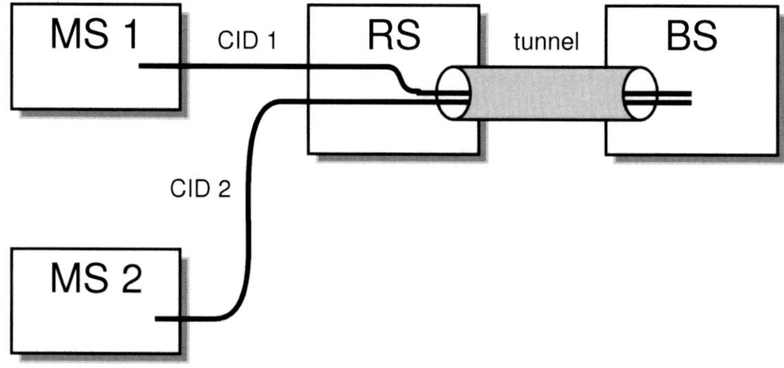

Figure 3.19: Tunnel connection with relay station

nels. Packets that pass the tunnel are encapsulated in a relay MAC PDU that contains the relay MAC header with the tunnel identifier. The protocol allows concatenation of multiple MAC PDUs that pass the same tunnel.

3.6.4 Bandwidth Request

Bandwidth requests in the relay enhanced system have to be handled by the RS and the BS sequentially.

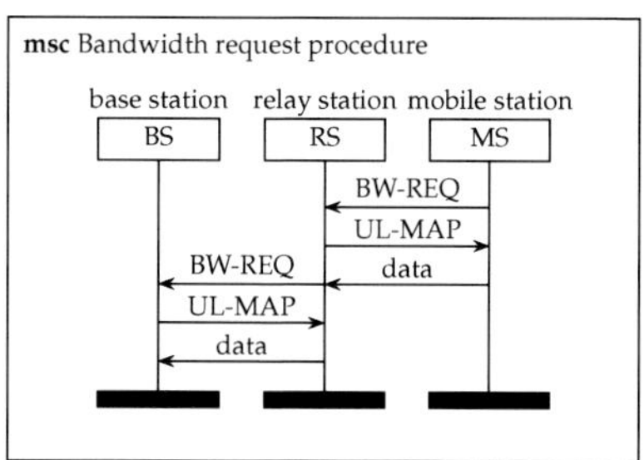

Figure 3.20: Bandwidth request procedure with relay station

Figure 3.20 shows how payload data is sent from the MS to the BS via the RS. The MS sends the bandwidth request to the RS. The RS allocates UL transmission opportunities according to the bandwidth request and the grant service that the MS has acquired in advance. The RS indicates the resource reservation for the MS in the UL-MAP. The MS sends the UL payload data to the RS. The RS sends a bandwidth request to the BS which sums all preceding bandwidth requests of subordinate MSs. The BS allocated UL transmission opportunities for the RS and indicates them in the UL-MAP. Finally the RS sends the payload data to the BS.

3.6.5 Radio Resource Control and Quality of Service

QoS support in multihop relay systems is more difficult to achieve compared to single-hop systems. Fairness in the multihop relay system has been investigated in (Gambiroza, Sadeghi, & Knightly, 2004) for the IEEE 802.11 system. (Chen, Xie, & Wu, 2009) presents an algorithm to assure fairness among users for relay enhanced systems but does not consider channel states that typically exists in mobile radio networks. (Lee, Narlikar, Pal, Wilfong, & Zhang, 2006) shows how to perform admission control to support QoS in the relay enhanced WiMAX system. (Bayan & Wan, 2010) shows how to integrate a QoS aware radio resource scheduler into the WiMAX multihop system.

Signaling and user data transmission between MS and BS is relayed by the RS. The BS controls the radio resource access of the RS for backhaul transmissions and for directly to the BS associated MSs. The radio resource scheduler of the BS has to assure that the QoS requirements of MSs served by RSs are met. As a result the BS allocates radio resources for the RS backhaul with priority. The RS controls QoS sensitive transmissions between RS and MS the same way like the BS.

RS enhanced systems require additional interference mitigation techniques since RSs may interfere each other. Like in conventional cellular radio networks, clustering and fractional frequency reuse can be applied. To achieve interference mitigation between RSs allocation of their radio resources has to be controlled centrally by the BS. On the other hand if RSs are placed well, RSs do not interfere each other and can serve their MSs simultaneously (Sambale, 2013).

CHAPTER 4

Models and Scenarios

Contents

ITU-R has defined a set of system performance requirements to be met by candidate systems to be part of the IMT-A system family (ITU-R, 2008b). This chapter gives a brief introduction of the model assumptions and the system requirements specified by the ITU-R for the IMT-A system evaluation.

4.1 Test Environments and Pathloss Models

The IMT-A evaluation guideline (ITU-R, 2008a) specify the test environments for any radio access technology proposal that intends to become part of the IMT-A system family. Since IMT-A systems are supposed to provide service in a wide range of scenarios, the evaluation guideline specifies four test environments covering the majority of radio access technology use cases. Scenarios are chosen such that typical deployments are modeled and critical system aspects can be investigated. Focus is on scenario specific performance measures like cell spectral efficiency and cell-edge user spectrum efficiency.

4.1.1 Indoor

The indoor test environment focuses on high user data throughput capacity inside buildings served by a small cell. The environment is characterized by high user density and heavy signal shadowing by walls. The test environment involves stationary and pedestrian mobile users in a floor of 16 rooms of $15\,\mathrm{m} \times 15\,\mathrm{m}$ and a hallway of $120\,\mathrm{m} \times 20\,\mathrm{m}$. The channel model of this scenario is called indoor hotspot (InH).

4.1.2 Micro-cellular

The micro-cellular test environment models continuous small cells in city centers and dense urban areas with high user densities and heavy user data load. The test environment includes pure outdoor as well as outdoor-to-indoor links. Base stations are assumed arranged in a regular hexagonal grid with antennas below rooftop. The channel model of this scenario is called urban micro (UMi) and is not considered in this work.

4.1.3 Base Coverage Urban

The base coverage urban test environment comprises large cells with continuous coverage in an urban environment. Base station antennas are assumed located above rooftops at a height of 25 m. Mobile stations are located outdoors at street level. Typical building heights are over four floors. The scenario comprises LoS as well as non line of sight (NLoS) signal propagation conditions. The channel model of this scenario is called urban macro (UMa).

The performance evaluation presented in Chapter 6 is based on the UMa model.

Table 4.1: Urban macro pathloss model parameters

Scenario	Pathloss [dB]	Applicability range
LoS	$l = 22\log_{10}(d) + 28 + 20\log_{10}(f_c)$	$10\,\mathrm{m} < d < d'_{BP}$
	$l = 40\log_{10}(d) + 7.8 - 18\log_{10}(h'_{\mathrm{BS}})$ $-18\log_{10}(h'_{\mathrm{MS}}) + 2\log_{10}(f_c)$	$d'_{BP} < d < 5000\,\mathrm{m}$
NLoS	$l = 161.04 - 7.1\log_{10}(W) + 7.5\log_{10}(h)$ $-(24.37 - 3.7(\frac{h}{h_{\mathrm{BS}}})^2)\log_{10}(h_{\mathrm{BS}})$ $+(43.42 - 3.1\log_{10}(h_{\mathrm{BS}}))(\log_{10}(d) - 3)$ $+20\log_{10}(f_c) - (3.2(\log_{10}(11.75h_{\mathrm{MS}}))^2$ $-4.97)$	$10\,\mathrm{m} < d < 5000\,\mathrm{m}$

Table 4.2: Urban macro pathloss model shadow fading parameters

Scenario	Shadow fading
LoS	$\sigma = 4$
NLoS	$\sigma = 6$

Table 4.1 shows the UMa pathloss parameters. Typical for this model is that the radio path has both, a LoS and a NLoS component with a distance dependent break point in between. Other parameters are: MS height $h_{\text{BS}} = 1.5\,\text{m}$, BS height $h_{\text{BS}} = 25\,\text{m}$, street width $W = 20\,\text{m}$, average building height $h = 20\,\text{m}$ and break-point distance $d'_{BP} = 4h'_{\text{BS}}h'_{\text{MS}}f_c/c$ where $c = 3 \cdot 10^8\,\text{m/s}$ is the vacuum light propagation velocity. The effective antenna heights h'_{BS} and h'_{MS} are calculated as shown in Eq. (4.1).

$$h'_{\text{BS}} = h_{\text{BS}} - 1.0 \quad \text{and} \quad h'_{\text{MS}} = h_{\text{MS}} - 1.0 \tag{4.1}$$

In addition the model assumes a time-invariant zero-mean log-normal distributed shadow fading with standard deviation σ according to Table 4.2.

$$p_{\text{LOS}}(d) = \min\left(\frac{18}{d}, 1\right)\left(1 - e^{\frac{-d}{63}}\right) + e^{\frac{-d}{63}} \tag{4.2}$$

As Eq. (4.2) shows, the LoS probability for a link depends on the distance between both stations and decreases with larger distances. Figure 4.1 shows the plot of Eq. (4.2) for distances up to 1000 m.

4.1.4 High-speed

The High-speed test environment focuses large cells and continuous coverage up to 10 km. Base station antennas are assumed located high above street level at 35 m. Mobile stations are located in vehicles and trains and have a velocity up to 120 km/h. The pathloss model of this scenario is called rural macro (RMa).

4.2 Antenna Model

The ITU-R has also published an antenna model for the BS in (ITU-R, 2008a) that provides a directional antenna characteris-

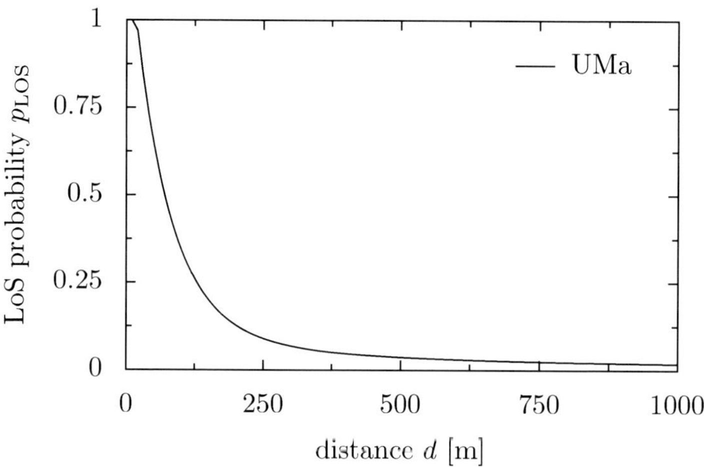

Figure 4.1: LoS probability vs distance

tics defined by elevation and azimuth components. Each cell of a BS is covered by an antenna of the same characteristics.

$$A_e(\theta) = -\min\left(12\left(\frac{\theta-\theta_t}{\theta_{\mathsf{3dB}}}\right)^2, A_m\right) \tag{4.3}$$

$A_e(\theta)$ defines the elevation component, θ_t the tilt of the antenna, $\theta_{\mathsf{3dB}} = 15°$ the elevation 3 dB value und $A_m = 20\,\text{dB}$ the maximum attenuation of the antenna.

$$A_a(\phi) = -\min\left(12\left(\frac{\phi}{\phi_{\mathsf{3dB}}}\right)^2, A_m\right) \tag{4.4}$$

$A_a(\phi)$ defines the azimuth component of the antenna characteristic. $\phi_{\mathsf{3dB}} = 70°$ defines the 3 dB value of the azimuth component. The overall antenna characteristics is given by Eq. (4.5).

$$A(\phi,\theta) = -\min\left[-(A_a(\phi) + A_e(\theta)), A_m\right] \tag{4.5}$$

4.3 PHY-Abstraction

Since detailed link level performance modeling in system level simulation would overload the simulation computer, link level performance is modeled separately, as usual. The published literature provides several options for physical layer abstractions. The exponential effective SINR mapping (EESM) is widely used since it is fast and easy to implement (Brueninghaus et al., 2005). Unfortunately the EESM does not provide information on error probability of segmented code-blocks dependent on the MCS used. Mutual information based SINR mapping (Brueninghaus et al., 2005) fills the gap and has become common practice for error modeling in system level simulation. (Srinivasan, 2009) introduces two MI based PHY-abstractions for evaluating the IEEE 802.16m system that allow to compute the mutual error probability of differently modulated code blocks. Here it is shown how to calculate the received bit mutual information rate (RBIR) as a means of physical layer abstraction.

The MI based SINR mapping is performed in two steps, see Fig. 4.2. The first step calculates the MI I_i from the SINR γ_i of all symbols that belong to a coded data block, shown on the left side of Fig. 4.2. This step is independent of the code used and depends on the modulation only. In the second step the coding model calculates the MI of the whole code block according to the MI metric and the block error rate (BLER) r_{BLE} according to the MI mapping.

4.3.1 Modulation Model

The first step in calculating the BLER of a packet transmitted by a number of segments is to calculate the MI of each segment whose size may vary from a few bits up to the size of a physical resource unit. In this work it is assumed a physical resource unit contains 18 sub-carriers and six symbol samples which

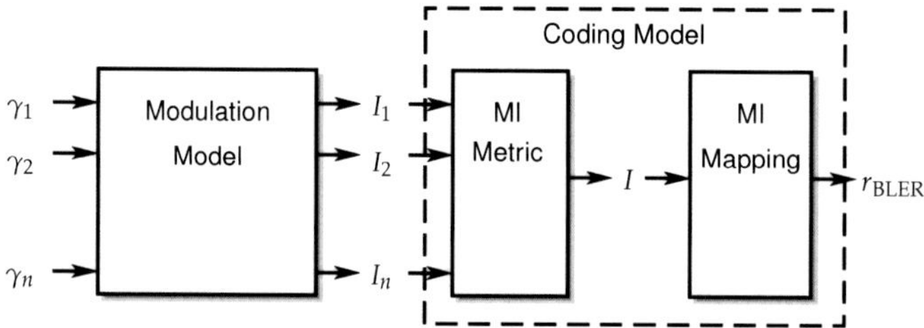

Figure 4.2: MIESM computation procedure

results in 108 OFDM symbols. For a given modulation scheme (Zheng et al., 2008) shows that under Gaussian interference MI only depends on the SINR. Derived from results of (Zheng et al., 2008, Section 4.2) it is possible to calculate lookup-tables mapping SINR to MI mapping for each modulation scheme considered.

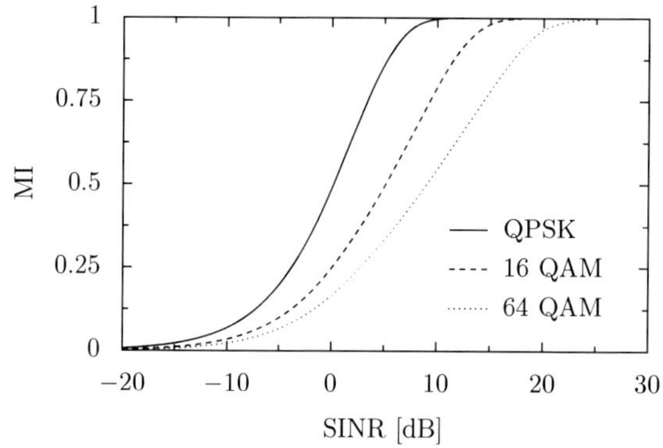

Figure 4.3: Mutual information vs SINR

Since computation of the MI needs to be fast for system level simulation, SINR to MI mapping is performed through a lookup-table according to Table A.1 shown in Appendix A. Figure 4.3 shows the resulting MI over SINR at the receiver for several schemes quadrature phase shift keying (QPSK), 16 quadrature amplitude modulation (QAM) and 64 QAM.

4.3.2 Coding Model

The MI metric evaluates the mutual information of the whole code-word from its segments. Equation (4.6) calculates MI I of the code block with size M from the segments I_i with size m_i.

$$I = \frac{1}{M} \sum_{i=1}^{n} m_i \cdot I_i \tag{4.6}$$

As Fig. 4.2 shows, the mutual information for each code block is mapped to the BLER r_{BLE}. Lookup-tables can be used to perform the mapping. Another alternative is to approximate the mapping function with a parametric function. A close fit to the performance curve can be given by a Gaussian cumulative model with two parameters b and c (Srinivasan, 2009).

$$y = \frac{1}{2}\left(1 - \operatorname{erf}\left(\frac{x-b}{\sqrt{2}c}\right)\right) \tag{4.7}$$

Since the mapping does not depend on the modulation, b and c can be provided for each channel code.

$$r_{\mathrm{BLE}}(m) = \frac{1}{2}\left(1 - \operatorname{erf}\left(\frac{x-b_{\mathrm{BCR}}}{\sqrt{2}c_{\mathrm{BCR}}}\right)\right) \tag{4.8}$$

For the performance analysis, b_{BCR} and c_{BCR} are given for all codes that are in use in Table 6.2 that are calculated from (Park, Lim, & Kim, 2009, Chapter 4).

4.4 Traffic Models

4.4.1 Voice Traffic

The voice traffic model is a two state Markov chain model known as Brady model (Brady, 1969) shown in Fig. 4.4. The active state represents a speaking user where the voice encoder generates VoIP data packets. The transition probability from active to inactive state is a. In the inactive state, the model creates silence insertion descriptor (SID) packets. The transition probability from inactive to active state is c.

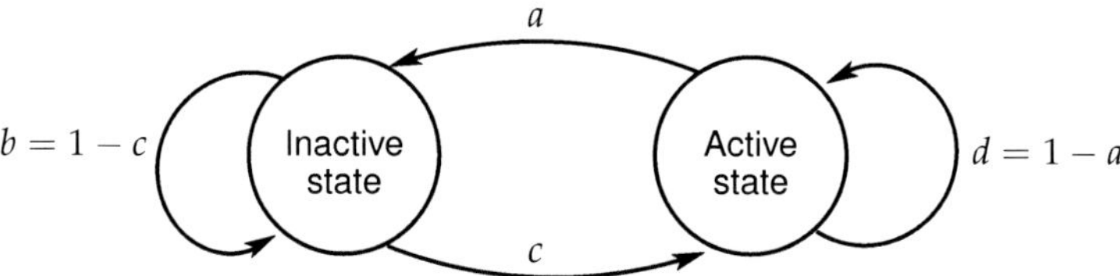

Figure 4.4: Brady voice activity model

Table 4.3 gives the parameters values for the adaptive multi-rate (AMR) codec commonly used in IMT-A systems (3GPP, 2012).

4.4.2 Payload Header Compression

Since a voice data packet contains a significant amount of redundant header information, the IEEE 802.16 system supports payload header suppression (PHS) for voice traffic. A common PHS scheme is robust header compression (ROHC) (Bormann et al., 2001), which requires an exclusive channel to transmit header compressed payload data. In the first packet of a talkspurt (after the state of the Brady model has changed from inactive to active), the sender transmits static context information for the

Table 4.3: AMR VoIP codec parameters

Parameter	Value	Description
Codec data rate	12.2 kbit/s	
Encoder frame length	20 ms	Inter-arrival time of encoder frame
Voice activity factor	50 % ($c = 0.01$, $d = 0.99$)	
SID payload	15 bytes	Inter-arrival time of SID frame: 160 ms
Protocol overhead	74 bit + padding bits	10 bit + padding bits (RTP-pre-header), 4 bytes (RTP/UDP/IP), 2 bytes (RLC), 16 bit (CRC)

payload transmitted. Subsequent packets that would contain the same context information instead carry compressed headers that can be decoded at the receiver without loss of information.

(Srinivasan, 2009) provides a model for both, ROHC supported and uncompressed voice traffic. Table 4.4 shows the parameters of the ROHC model.

For VoIP users, packet delay and packet loss are crucial parameters. A mobile user is assumed satisfied with the VoIP service, if 98% of the packets are delivered within a delay of 50 ms. The VoIP packet delay is measured as the overall latency starting from packet transmission at the transmitter to the packet decoder at the receiver station. VoIP packets delayed more than 50 ms are considered lost in the simulation study. Packet delay in the core network is not included here. The system VoIP capacity is the number of MSs that can be served with an outage of less than 2%.

Table 4.4: AMR voice traffic model parameters

	without compression	with compression
RTP header	96 bits	
UDP header	64 bits	
IPv4 header	160 bits	
ROHC header		24 bits
payload headers sum	320 bits	24 bits
802.16 generic MAC header	48 bits	48 bits
802.16 CRC for HARQ	16 bits	16 bits

4.4.3 Full Buffer

The full buffer model of a station generates fixed size data packets thereby the highest possible load to a communication system. The model represents a data packet buffer that never exhausts of packets. This model is used to measure the maximum possible throughput capacity of a system, see Section 6.6.2.

4.5 Association of MS to the Wireless Network

In the process of network entry, a MS decides which BS or RS is adequate to provide the desired service. There are several criteria that can be used to choose among the potential serving BS. All of them use the signal strength of the control channel of the candidate BSs or RSs.

Lowest pathloss

The MS chooses the BS or RS that offers the lowest pathloss. The pathloss can easily be calculated by subtracting the received carrier power of the broadcast channel by the transmit power communicated by the BS.

Highest carrier power
The MS chooses the candidate station with the strongest received signal strength. This method is equal to *lowest pathloss* method if all BSs transmit with the same transmit power.

Highest SINR
The MS chooses the candidate station that offers the highest SINR. This allows the MS to take interference power into account.

4.6 Relay Placement

The number and the position of relays in a cell is crucial to achieve a benefit in system performance. On the one hand, RSs have to maintain a link with high capacity to the super-ordinate BS or super-ordinate RS (not considered here). On the other hand, the RS has to cover regions of the cell, that the BS can hardly serve. Sambale (Sambale & Walke, 2012b) has shown how optimum RS positions in the IMT-A reference scenarios can be found through simulated annealing that maximize the system performance while keeping the number of RS low. In this work we take the locations found to be optimum in (Sambale & Walke, 2012b).

4.7 Coordination of Relay Stations

With the integration of RS in the cellular system, the question arises if multiple RSs may operate simultaneously in space division multiplex (SDM) mode for DL and UL. RSs may even serve their associated MSs in parallel to the BS serving its associated

MSs. The simultaneous operation may result in a gain of cell capacity if the RS's mutual interference is negligible. On the other hand, if mutual interference of RSs prevents an acceptable SINR, RS need to operate on exclusive resources, at least partially.

Recent studies in relay enhanced cellular radio networks that rely on the IMT-A evaluation guidelines have shown that the SDM mode of operation of RSs is beneficial (Sambale & Walke, 2012b). If RSs are placed carefully, mutual interference can be minimized and the system shows the best capacity gain with SDM operation.

4.8 Simulation Methodology

To achieve statistically confident results simulations are performed several times according to the batch means method (Cox & Lewis, 1966) with the same system parameters but with a different initial seed of the random number generator. As a result the MSs positions in cells differ for each simulation run. Each run or *drop* runs for a simulation time of 20 s where the last 15 s are used to evaluate performance parameter values by measurement. The first 5 s give the system the opportunity to reach a steady state without taking measurements.

CHAPTER 5

The openWNS Simulator Platform

Contents

Network simulation is an established method to develop, investigate and evaluate network protocols and device features like transmission power and antenna characteristics. Developments in computer technologies allows to simulate large wireless network on system level in detail. The high computing power available even allows to combine system level network simulators to study details of complex system models.

5.1 Event driven Protocol Simulators

The following section presents the most common event driven system level simulators in research and compares their features.

5.1.1 OPNET

OPNET Technologies Ltd. offers a commercial simulation environment and network modeling tools for a wide range of communication protocols. OPNET is a discrete stochastic event-driven simulator platform for system level simulation. It comprises modeling tools, simulators and evaluation tools. With the Modeler Wireless Suite® (OPNET Technologies, 2013), OPNET provides additional components to simulate wireless protocols like Global System for Mobile Communication (GSM), IEEE 802.16 Mesh, etc. The simulator platform allows for implementing and evaluating proprietary protocols and network configurations. OPNET offers a graphical editor that allows to build network topology and protocol stacks from the application layer to the PHY layer. It follows a hierarchical architecture that maintains a network of nodes containing processes that define protocols.

5.1.2 OMNeT++

OMNeT++ is an discrete event simulation platform available free of charge for academic and educational use. Its primary application area is the simulation of communication networks, but because of its generic and flexible architecture, is successfully used in other areas like the simulation of complex IT systems, queuing networks or hardware architectures as well (OMNeT++, 2014).

OMNeT++ provides a software architecture to support system components implemented in C++ and can be assembled using the high-level network description (NED) language. One of the strengths of OMNeT++ is its extensive graphical user interface (GUI) support. This makes it easy for beginners to achieve results early. In addition, the GUI helps the user to analyze the protocol and to find bugs.

5.1.3 NS-3

NS-3 is an open source discrete-event network simulator for Internet systems, targeted primarily for research and educational use. ns-3 is free software, licensed under the GNU GPLv2 license, and is publicly available for research, development, and use (Antipolis, 2007-2013).

Basically, the NS-3 is a C++ library which provides a set of network models. The network models can be connected in a dynamical way, which enables to investigate arbitrary network and protocol setups. NS-3 uses Python bindings to provide a configuration framework for modular protocol stacks and to setup the simulation scenario as a whole.

Since NS-3 is a popular tool in the scientific community, there are a lot of additional modules that provide special protocols like transmission control protocol (TCP), Internet protocol version 4 (IPv4) and Internet protocol version 6 (IPv6), routing, peer-to-peer protocols and more. The main focus of the NS-3 simulator are network and transport protocols. As a result link-layer protocols are rare and less detailed.

5.1.4 Vienna LTE Simulator

The Vienna LTE simulator is an open source combined link- and system-level simulator implemented in MATLAB available under academic non-commercial use license (Mehlführer et al., 2011). It offers bit-accurate link level simulations and PHY-model-based system level simulations for LTE. The system level simulator focus on radio resource allocation and scheduling, interference management and network optimization. The intention of the authors is to provide a simulator that can be used to reproduce published research results in the context of LTE.

5.1.5 openWNS Simulator

The reuse of protocol elements requires a common interface between protocol units. (Schinnenburg, Pabst, Klagges, & Walke, 2007) shows a concept to interconnect protocol elements that are divided into FUs.

The openWNS is an open source highly modular simulator for performance evaluation of mobile radio networks. It is written in C++ and provides Python bindings to support run-time configuration. The simulator configuration can be defined in a Python script and the simulator runs the configuration without recompilation. openWNS provides a binary executable simulator application that includes a Python interpreter to import simulator configuration[1]. It allows for stochastic event driven system level simulation based on prototypical protocol implementations.

The goal of the openWNS platform is to provide a simulator which is suitable for various kinds of performance evaluation of wireless and wired networks. A point of special interest is the support of cellular mobile radio networks. The key idea is to have a flexible framework which allows the researcher to adapt

[1] It has been developed at ComNets, RWTH Aachen University from 2004 on.

the grade of complexity and accuracy of the simulation model to his needs for the respective investigation.

The second major goal is a simulator which supports the researcher in testing his implementations in order to help him find errors at early stages during his investigation. Therefore it provides comfortable testing frameworks at different levels of testing, from unit testing to system testing. These frameworks have been carefully tailored to an event-driven, stochastic simulation system (Bültmann, Muehleisen, Klagges, & Schinnenburg, 2009).

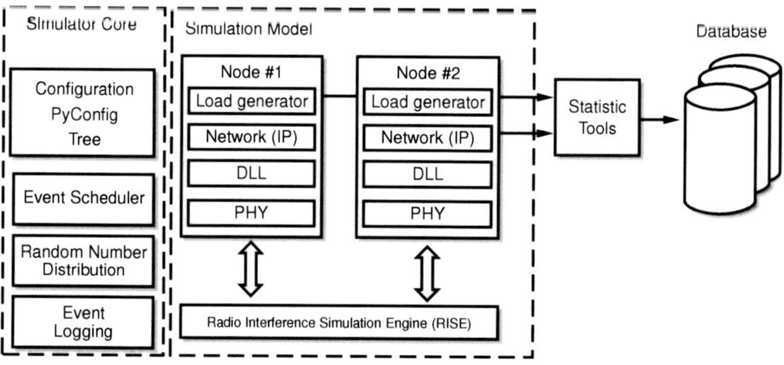

Figure 5.1: Simulator Architecture (Bültmann et al., 2009)

The majority of the core simulator components is implemented in the openWNS library. In addition to protocol implementations, it contains common software design patterns like observers, singletons and more that can easily be used in plugins or extensions of the simulator platform.

Figure 5.1 shows the openWNS architecture dividing the platform into the simulator core components like the Python configuration utilities, the event scheduler, random number generator and event logging and in the simulation model. The center of the figure shows the simulation model by example of two nodes each with its protocol stack. Nodes are connected by the common Radio Interference Simulation Engine (RISE) layer that allows to deliver PDUs via a simulated radio interface. On the right side of the figure shows the statistics tool set that processes statistical data for later delivery to the data base. The openWNS is free software and licensed under the GNU Lesser General Public License (LGPL). It can be obtained from (openWNS, 2004-2013). Table 5.1 summarizes the features of the system level simulators mentioned.

Table 5.1: Simulator feature list

	OPNET	**OMNeT++**	**NS-3**	**LTE Simulator**	**openWNS**
open source	✗	✗[a]	✓	✓	✓
graphical editor	✓	✗	✗	✗	✗
object oriented	✓	✓	✓	✗	✓
scripting language	✗	NED	Python	Matlab	Python
graphical eval.	✓	✓	✓	✓	✓
grid computing	✓	✓	✓	✗	✓
wireless protocols	IEEE 802.11, WiMAX, LTE, UMTS...	LTE, IEEE 802.11	IEEE 802.11, WiMAX, LTE	LTE	WiMAX, LTE, IEEE 802.11
operating system	Windows	Windows, Linux, Mac OS X	Linux	Matlab	Linux, Windows

[a] free for academic use

5.2 openWNS in Detail

In this work openWNS has been used and extended to be able to evaluate the performance of relay enhanced cells. This is the reason why the simulator is described in more detail compared to the other simulators mentioned. Emerging air interfaces have shown that different protocols mostly contain similar functions.

As a result it is beneficial to reuse functions in different protocols. First approaches to design a modular protocol stack that reuses protocol functions can be found in (Siebert, 2000; Siebert & Walke, 2001). Berlemann et al. extended the work to design a flexible protocol stack for multi-mode operations that support multiple air interfaces (Berlemann, Pabst, Schinnenburg, & Walke, 2005). The openWNS follows this principle throughout.

5.2.1 Run-time Configuration

The configuration of the simulator is specified in Python language which is interpreted by the simulator software at run-time. For each component, model or service, there exists an equivalent representation in the Python scripting language. Similar to the object oriented design of the simulator, the configuration contains node, components and models. At the beginning of the simulation, the simulator iterates through the simulator configuration that exists in the Python-domain and creates equivalent C++ objects. For this, the simulator uses static object factories that are able to identify the object in the Python-domain by a keyword. The factory provides a creator object that is statically bound to the keyword. The creator object creates an instance of the desired type and returns the object to the questioner. If the object contains nested objects, the factory creates nested objects recursively.

The flexible run-time configuration requires that all classes in the C++ domain are registered with a name at the factory that matches the keyword in the Python-domain.

5.2.2 Statistic Evaluation

The statistic evaluation system of the openWNS provides a framework that can take measurements at arbitrary locations in the protocol stack. It offers filters and functions to sort measurements by contexts and accumulate measurements in evaluation sinks.

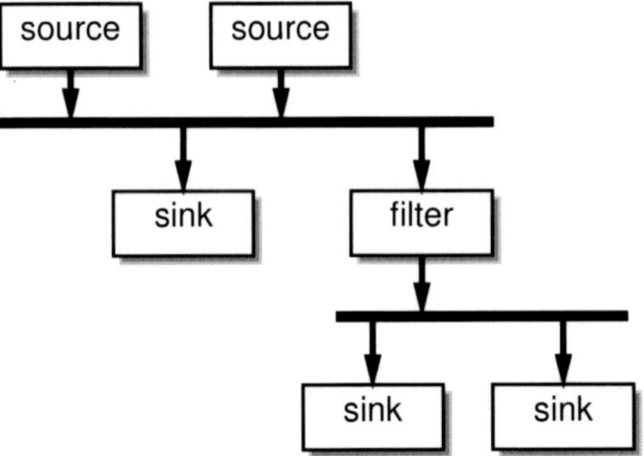

Figure 5.2: Probe bus

The central element of the statistic evaluation framework is the *probe bus*. The *probe bus* connects all components of the probing system for the whole simulator. Probe sources send measurement reports that are distributed by the probe bus to all measurement sinks or subordinate probe buses. Figure 5.2 shows a block diagram of a probe bus example setup with two sources. The top-level probe bus distributes the measurement messages to a sink and a filter. The filter forwards selected measurement messages according to a freely configurable set of rules to the second probe bus that distributes the measurement messages to two sinks.

The statistics framework offers a continuous online data processing at a constant memory consumption, even for long simulation runs. The online data processing does not result in log-files that grow over time and contain all data that occur in the simulator. Instead the user defines data processing rules a priori that drop data of little importance. The advantage of the statistical processing of measurements over unprocessed measuring is the compression of measurement results.

5.2.2.1 Data Sources

Measurement data is almost always related to a specific context. For example, buffer fill levels are related to a specific service flow or interference power values are related to a specific station. The statistic evaluation framework binds a measurement result to a context that contains additional information about the source from where the measurement stems. The context contains a list of context descriptors. A name of the context, like *station type* or *service flow* and the related integer value. Since the context of the measurement is available on the probe bus, filters may access the context and may take actions like dropping or sorting according to the context values.

5.2.2.2 Data Sinks

In general, sinks are evaluation components for statistic measurement results. Sinks are reached via the probe bus. The simulator offers several data sinks that provide different functions as explained in the following.

Moment

The moment sink provides statistical moments like mean, variance, standard deviation, second and third moments, minimum, maximum and the number of samples used to calculate a moment.

Probability Density Function

The probability density function (PDF)-sink provides histogram, PDF, cumulative distribution function (CDF) and complementary cumulative distribution function (CCDF). The PDF-sink also provides percentiles like P01, P05, P50, P95, and P99. The PDF data sink requires specification of a range of expected values. The range is given by the bounds of the range, i.e. expected minimum and maximum are to specify during configuration of the PDF data sink. Additionally, the number of bins for the histogram needs to be specified.

5.2.3 Testing Framework

Correctness of the implemented models is of paramount importance for reliability of simulator results. The platform offers two options to check model and code correctness.

5.2.3.1 Unit Tests

Unit testing serves to test a well-defined function of a delimited unit of software in general (IEEE, 1986). In the context of openWNS the unit may be a service, a FU or other components of the simulator platform that is implemented in a class. The unit test creates an environment that encapsulates the device under test. The testing environment that is also called test-fixture feed the unit with well-defined input data and checks if the expected results occur at the output. The unit test does not test the internal realization or the internal state of the unit. This allows for a free implementation of the unit without any constraints on the internal function.

Test fixtures of all modules and components are registered in a global list of test fixtures that is also called global test suite. If the openWNS simulator is started with the test-flag, the simulator executes all registered tests of the global test suite and quits. The simulator shows the test results on the screen and highlights tests that failed. Unit tests are the basis of FU tests, see Section 5.2.4.5.

5.2.3.2 System Tests

Interaction of units in a protocol stack or even in a mobile radio network system can hardly be tested in the unit test framework, since it only allows to test a single or a small amount of units. To check the correctness of a protocol stack, the simulator platform offers system tests that involve nodes with fully equipped protocol stacks, channel models and traffic load. The simulator starts with the setup of the test scenario, performs the simulation run and performs the system test checks. Like with unit tests the system test checks if a system provides the expected output under well-defined circumstances and thereby the conformance of the protocol implementation to what is expected as output to a given input. If the system test does not meet the expectations, the failure is highlighted on the screen or in the simulator log files.

5.2.4 Functional Units

With the investigation of multiple wireless and mobile radio systems in a single simulator, the idea to share functions between the protocol layers arises. A common interface for atomic protocol functions embedded in units called functional unit (FU) enables the simulator to share functions in multiple protocol layers. Interface definition, flow control procedures and software architecture are summarized in the FU concept partly developed by the author of this thesis (Schinnenburg, Debus, Otyakmaz, Berlemann, & Pabst, 2005; Schinnenburg et al., 2007).

In the context of the FU concept, implementation of a protocol is a decomposition of complex functions into basic FUs. FUs are connected and exchange data by means of compounds. Each FU contributes a command to the set of commands and has a cohesive responsibility in terms of protocol functionality. A common FU interface assures the interoperability of all FUs.

The approach to decompose a protocol layer into atomic protocol FUs is known from the literature and has proven to be beneficial to deploy new mobile devices in short update cycles (Schöler & Müller-Schloer, 2004).

5.2.4.1 Data Handling

In the context of the openWNS protocol layers are called components (Schinnenburg et al., 2005, 2007). Components contain the FUN that is constructed of connected FUs. Like a layer adds PCI to PDUs, a FU adds control data to a data frame. The FU concept defines the PDU as a *compound* of the SDU and a pool of commands. Each command is the dedicated data of one FU and is accessible only by the FU.

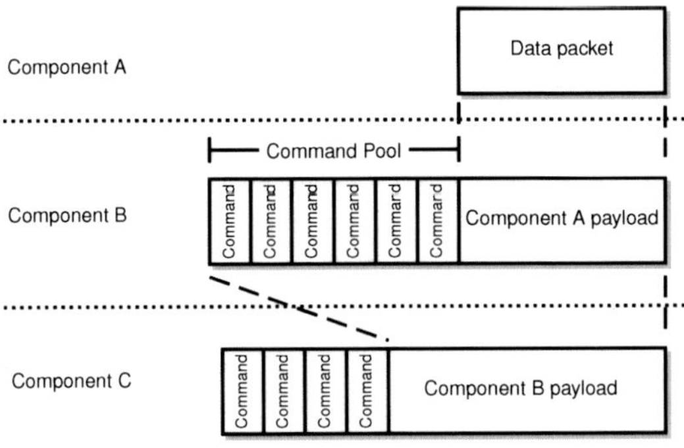

Figure 5.3: Compound and commands

Figure 5.3 shows a compound that contains the payload data and the pool of commands of multiple components. Component B encapsulates the data packet of component A and adds the command pool for its FUs. Command pool and payload data form the compound of component B. The compound of component B can be the payload data of a third optional component C. Components that create compounds by itself like load generators do not need to encapsulate payload data. The data model of the FU concept is similar to the ISO/OSI reference data model described in Section 3.1.

The packet data size is calculated by the size of the payload data and the size of the commands. As the compound travels through the FUN some commands may be still empty while others have been filled. To keep track of the used commands, FUs activate their command as soon as the compound passes the FU. Furthermore, the FU defines the size of the command by its content. The size of the compounds is the sum of the sizes of activated commands and the size of the payload.

5.2.4.2 Flow Control

In a real-world protocol stack, the offered traffic may exceed the capacity of the physical layer. In such a case, the protocol uses buffers to prevent packets from flooding the physical layer. In practice, the request to send a packet blocks the execution of the requester as long as the physical layer cannot provide the requested resource. The execution of the request continues as soon as the physical layer provides sufficient resources to serve the request.

The software concept needs to implement an equivalent flow control function that provides the following functions:

- Block service requests for outgoing traffic if the request exceeds the capacity or if the service provider is not able to handle the request for any other reason.
- Notify service users about free capacity as soon as it is available again.

The concept provides the functions by the *isAccepting*, *sendData* and *wakeup* calls of the FU interface. Figure 5.4 shows the flow control functions of three FUs for outgoing traffic. First all

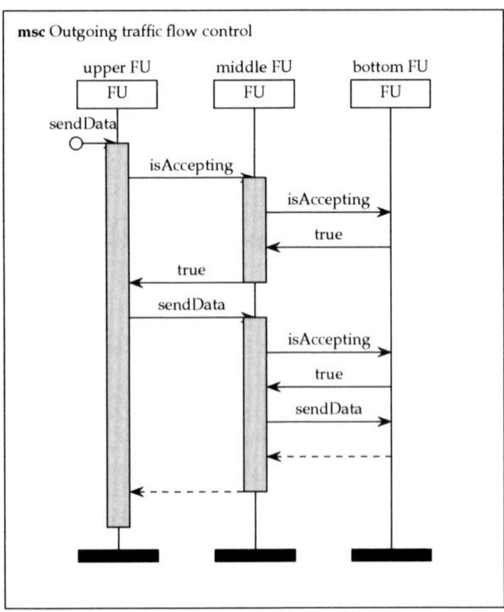

Figure 5.4: Flow control for outgoing traffic

FUs are in a sleep state. As soon as the upper FU wakes up by an external event it calls the middle FU by *isAccepting* request to ask for permission to send a compound. Since the middle FU cannot decide this, it forwards the request to the bottom FU. If the bottom FU accepts the compound by responding *true*, the middle FU forwards the reply to the upper FU.

Then the upper FU in return sends the data to the middle FU. Since the middle FU is not allowed to store the result from the first *isAccepting* request it issues a second *isAccepting* request to the bottom FU. Finally, the middle FU sends the data to the bottom FU.

Flow control events can also be triggered from any FU. For example, the physical layer observes free radio resources and is able to transmit data. In such a case the lower FU emits a *wakeup* call to the upper FU. Upon the *wakeup* call, the upper FU starts the *isAccepting/sendData* process as shown above. Figure 5.5 shows the message sequence chart of the *wakeup*-initiated flow

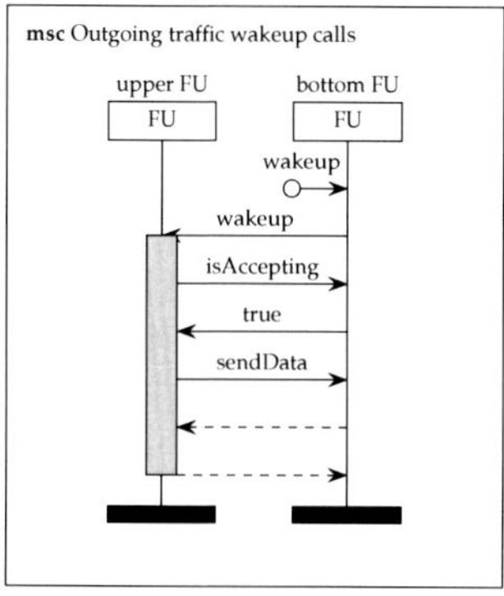

Figure 5.5: Wakeup calls for outgoing traffic

control process. If the upper FU does not send compounds by itself, it forwards the *wakeup* call to the next upper FU.

The order in which a FU wakes units from its receptors affects the behavior of the compound flow. FUs called first have a higher chance of sending their compounds. To make flow control fair, the receptor strategy should wake the FUs according a round robin process. For three FUs named A, B and C a fair wakeup strategy would be ABC, BCA, CAB, ABC etc. The standard receptor that allows to connect multiple FUs follows this rule.

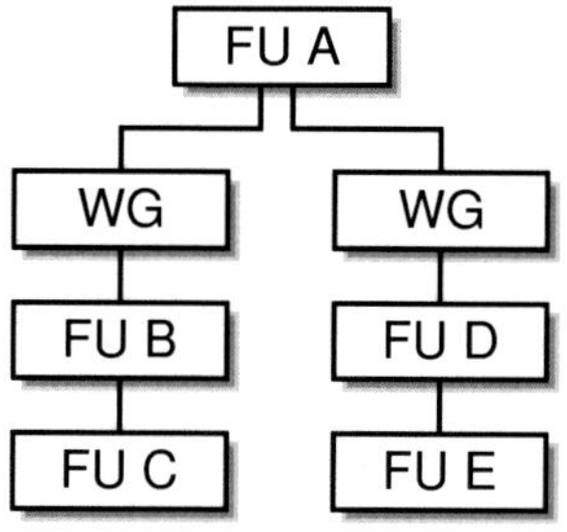

Figure 5.6: Fuctional Unit Network with Wakeup Gate

The flow control process bears the risk of needless excessive *isAccepting* calls if the chain of connected FUs is split into several paths. Figure 5.6 shows a FUN with a path that splits below FU A. If FU C receives a wakeup call from below, the call goes through FU B up to FU A. The flow control process foresees that FU A invokes isAccepting calls to the next FUs, which includes FU B and FU D even if FU D has not invoked the wakeup request to FU A. To prevent the needless isAccepting call chain in FU D, the paths are protected by two wakeup gates (WGs) that block the isAccepting call, if there has not been a wakeup request passing the WG before.

5.2.4.3 Relaying

Relaying is essential in current communication protocols. The concept needs to allow resending compounds that have been received in advance. Resending compound is not trivial since the compound contains activated commands. Activating of previously activated commands is prohibited. Instead the concept foresees to create a partial copy of the compound that contains the activated commands of the original compound up to a specified limit.

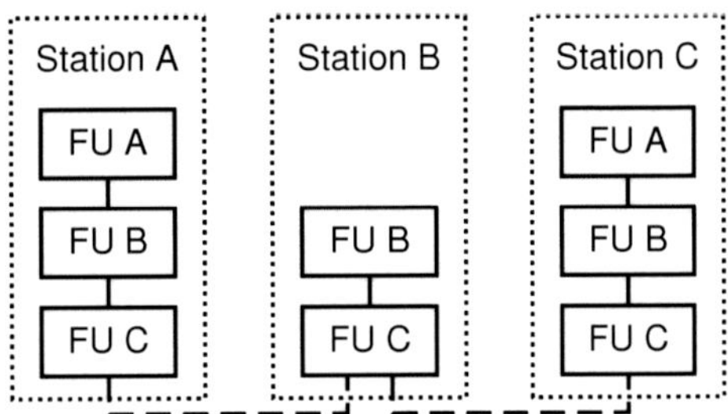

Figure 5.7: Relaying with Functional Units

Figure 5.7 shows the FUN of three stations. If FU A in station A sends a compound, the compound travels through FU B and C. Station A transmits the compound to station B and FU C of station B receives the compound. It sends the compound up to FU B. FU B creates a partial copy of the compound that contains the activated command of FU A but without the commands of FU B and C. FU B activates its command in the copied compound and sends it downwards. Since the compound does not contain any activated command of FU C, FU C in station B activates its command as usual. When the compound arrives in FU A in station C, the command contains the original information of FU A in station A.

In the case, the copy of the compound should not contain activated commands from FUs that are located above the FU that creates the partial copy, the FUN contains a *marker* FU that marks its command in the chain of activated commands. The partial copy contains the activated commands up to the marked command.

5.2.4.4 Rich Connections

The functional unit concept divides the FU into five aspects that provide defined functions. The five aspects of the FU are: *command type specifier*, *compound handler*, *connector*, *receptor* and *deliverer* (Schinnenburg et al., 2007). Connector, receptor and deliverer take a major role for flow control in the FUN. If a FU is connected to multiple FUs the instance of the connector, receptor or deliverer chooses among the available FU according to its strategy respectively to send a compound or to wake the next FU.

Although *connectors*, *deliverers* and *connectors* allow to connect multiple FUs they do not allow the compound handler to choose among connected FUs. To overcome the limitation, the concept has been extended in this work by named ports that either can be used to address a specifically connected FU to send compounds or to identify the origin of a received compound. A connection via a named port is called rich connection.

Figure 5.8 shows an extended FU that provides two ports, *Port A* and *Port B* on the top of the FU. As the figure shows,

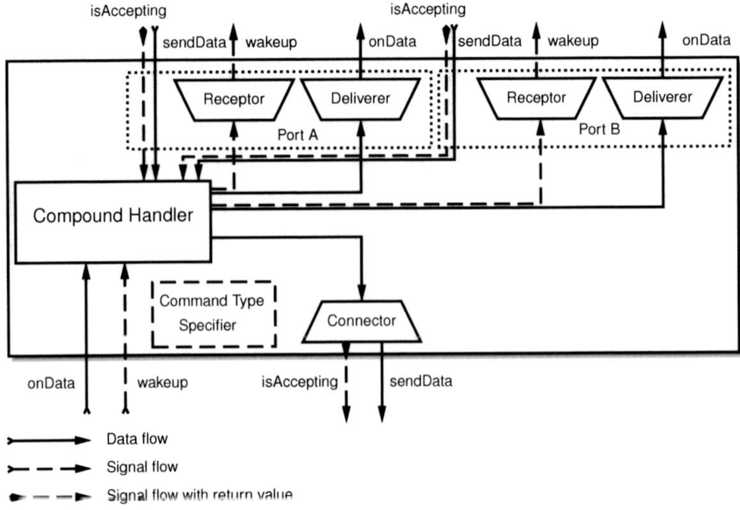

Figure 5.8: Functional Unit with Rich Connections

the deliverer and the receptor entities exist for each port. Each port provides the *wakeup* and *onData* interface through deliverer and receptor entities. For incoming compounds, the compound handler can choose between the ports to send the compound up to the next FU. Furthermore it can choose the port to send *wakeup* signals. Additionally, the compound handler provides separated *isAccepting* and *sendData* interfaces for both ports. The rich connection extension takes an important role in QoS supported scheduling of the WiMAC protocol stack.

5.2.4.5 Functional Unit Testing

With the commonly define interface of FUs it is easy to implement a standardized FU specific test fixture that allows to perform FU tests without the need to create the test environment for each FU. The platform offers additional helper FUs that can be used for common tests. Figure 5.9 shows the test fixture that take a FU under test.

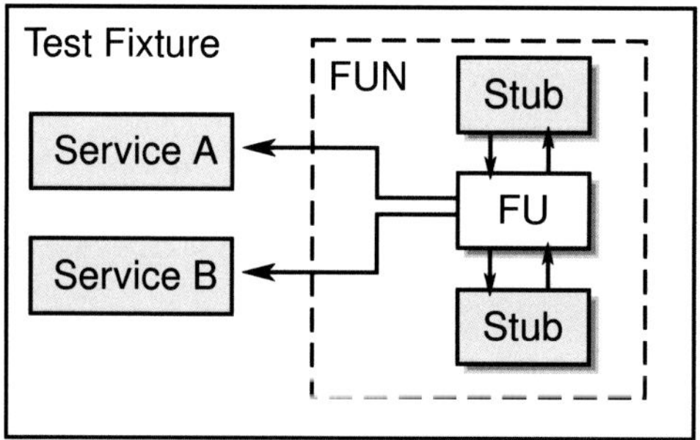

Figure 5.9: Functional Unit Test Fixture

The *Stub* FU is a helper FU that supports the functional unit test fixture to feed and observe a tested FU. The stub is a FU that contains counter that observe the number of compounds that have been sent through the *sendData* request and the number of compounds that have been received through the *onData* indication.

Two stubs enclose the tested FU and can take measurements about the compound flow. The test fixture provides a FUN that holds all FUs. Furthermore, the test fixture instantiates all needed protocol services. With the stubs as helping FUs the test fixture executes all tests that are defined for the specific FU under test. The stubs feed the FU with compounds of specified size and attributes. After the FU has processed the compounds the test checks if the compounds have reached the receiving stub and if the compounds have been processed correctly.

5.3 WiMAC - 802.16 Module

The openWNS provides a large set of design patterns that support the development of customized communication protocols and wireless systems. In addition to the given components of the openWNS, the WiMAC module provides IEEE 802.16 specific components like DL- and UL-MAP handlers, connection classifiers, frame building components and frame schedulers. The combination of standard protocol components and WiMAC specific components realizes the 802.16 protocol in the openWNS. The following sections show the components of the WiMAC that have been developed as part of this thesis.

5.3.1 Radio Resource Scheduler

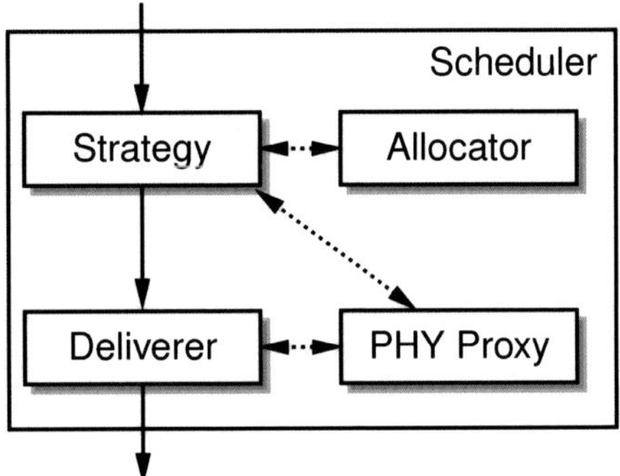

Figure 5.10: Scheduler Structure Diagram

The radio resource scheduler provides two major functions. The scheduler implements strategies that manage priorities among users that take the channel quality of the users into account. Furthermore the resource scheduler allocates resources for jobs that fit a given criterion best. The criterion may be the expected SINR or the expected capacity of a resource unit.

Figure 5.10 shows the internal structure of the radio resource scheduler. The scheduler is separated into four components allowing a flexible reuse of the scheduler components in different physical layers and even in different higher layers. The modular structure of the resource scheduler supports the investigation of different resource allocation strategies as well as channel quality oriented adaptive packet scheduling.

5.3.1.1 Resource Unit

The resource unit (RU) is the smallest radio resource unit the scheduler can handle. The scheduler allocates a resource unit to a single user. If the scheduler allows for user multiplexing on a single resource unit, the resource unit can be allocated for multiple users. This can be useful if the channel connecting to multiple users has sufficient good quality so that the same MCS can be used on the same RU. Depending on the PHY layer configuration, the PHY proxy translates the abstract resource unit to a time-slot and sub-channel, see Section 5.3.1.4.

5.3.1.2 Job

The abstract *job* specifies the task the scheduler strategy has to process. The job is defined by a size and a destination. In general the job is the abstract equivalent to a MAC PDU that has to be processed by the scheduler.

5.3.1.3 Strategy

The *strategy* defines the selection strategy of the scheduler job and is also called discipline. If more than one job is waiting for delivery, the strategy defines the order of processing among the waiting jobs. The simplest scheduler strategy is the first come first serve (FCFS) strategy.

5.3.1.4 PHY Proxy

The *PHY proxy* maps RUs of the scheduler domain on PRUs carried in time slots on sub-channels of the system. The mapping depends on the physical layer of the system. There is a one-to-one mapping between resource unit IDs and time-slot-/sub-channel numbers.

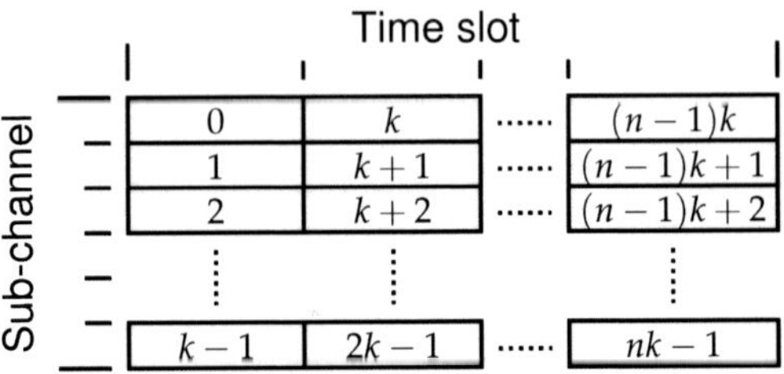

Figure 5.11: Resource unit mapping

Figure 5.11 shows a radio resource mapping on a radio channel with k sub-channels and n time slots. The resource unit number increases with the number of sub-channels and wraps with the number of the time-slot. The mapping function of the resource number l is given in Eq. (5.1).

$$l = (i-1) \cdot k + j - 1 \quad \forall i \in \{1 \dots n\}, j \in \{1 \dots k\} \tag{5.1}$$

The PHY-Proxy allows the scheduler to operate on a one-dimensional space without the need to handle slots or sub-channels.

In addition, the PHY proxy provides the scheduler with radio resource quality information. It offers the expected channel state to a given receiver for a specific radio resource unit. With this information the strategy decides which and how many resource units are to be allocated for a scheduler job.

5.3.1.5 Allocator

The *allocator* defines the type of the resource allocation. UL- and DL-scheduler need to allocate resource units in a different way. DL scheduler occupies a resource unit with a MAC PDU or a segment of a MAC PDU. The UL scheduler blocks the resource unit but does not occupy the unit with a radio transmission.

5.3.1.6 Deliverer

The *deliverer* implements the gate to lower FUs. It delivers the final resource schedule to the PHY layer but still allows other FUs to modify the compounds if needed.

5.3.2 Schedule processing

DL and UL resource scheduling are similar and are usually performed in the BS. For UL, the scheduler performs radio resource scheduling with information based on bandwidth requests that represent the traffic load in the MSs.

Figure 5.12 shows the flow chart of the scheduling process. The schedule process starts with a trigger that an external component sends to the scheduler. The strategy chooses a job that has to be processed first. The strategy acquires the channel state estimation from the PHY proxy and sends allocation requests to the allocator according to the channel state information and the allocation strategy until the job has been completely scheduled on the radio resource. The process repeats until all waiting jobs have been scheduled or all radio resources of the radio frame are occupied. Finally the scheduler hands the schedule over to the deliverer that creates transmissions on each physical resource according to the resource allocations taken by the strategy.

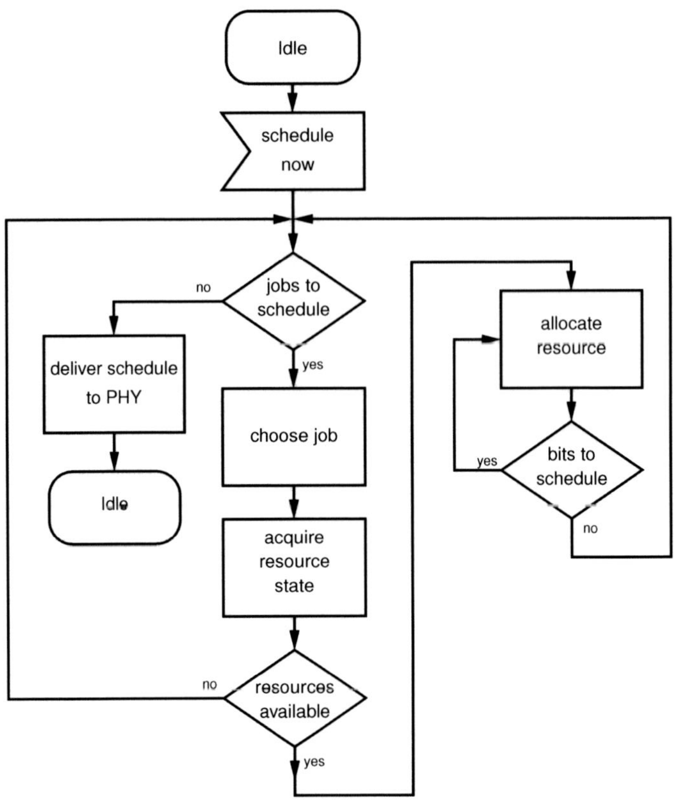

Figure 5.12: Schedule process

5.3.3 Channel State Estimation

Channel state knowledge is essential to be able by the scheduler to apply link adaptation. The WiMAC offers a protocol service for channel state prediction that allows for configuration of the prediction strategy. All of the channel state estimation strategies use channel state information (CSI) sent from the MS to the BS/RS. A strategy defines how the channel state information is used for the prediction. Channel state estimation is performed per resource unit and per intended receiver station. As a result the channel estimation differs for each resource unit depending on current pathloss and interference.

5.3.3.1 Lower Bound Strategy

The *lower bound* estimator always returns the lowest measured CSI value as prediction. In a synchronized system SINR for data frames is equal or higher than the lowest expected SINR measured in the broadcast channel.

5.3.3.2 Kalman Filter Strategy

Leung (Leung, 1999) has shown that the Kalman-Filter is a valuable predictor for interference power in a time division multiple access (TDMA) based cellular packet radio network. In (Leung, 1999) the predicted interference power I_n is calculated as:

$$I_n = I_{n-1} + F_n \tag{5.2}$$

where F_n is the fluctuation of interference power for slot n due to start and stop of transmissions. The measured interference power Z_n is given by:

$$Z_n = I_n + E_n \tag{5.3}$$

where E_n denotes the measurement noise. For a proper channel estimation the interference needs to be normal distributed. With a sufficient large amount of interfering stations, the central limit theorem states that the interference power at the receiver follows a normal distribution as long as the interfering stations are non-correlated.

5.3.4 Frame Builder Framework

The frame builder framework has been developed in this thesis to implement periodically recurring radio frames of the IEEE 802.16m protocol. The framework offers means to define logically separated phases that do not overlap in time. Each phase is solely controlled by a *compound collector* which is virtually a FU with special abilities shown in Section 5.3.4.3.

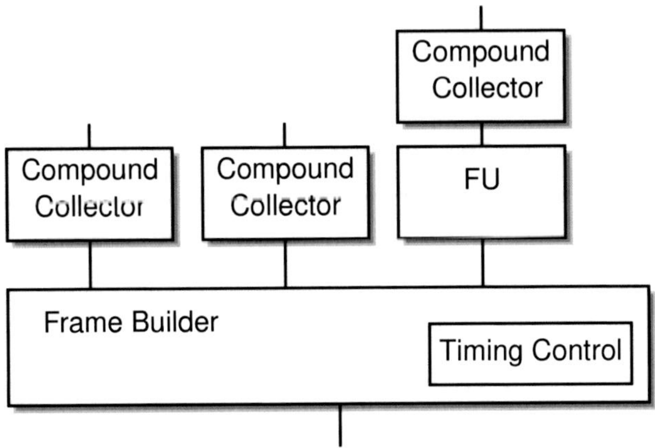

Figure 5.13: Frame Configuration Framework

Figure 5.13 shows a frame builder connected to several compound collectors. Since it provides the FU interfaces, it can be integrated into a FUN like a usual FU. The following describes the components of the frame builder framework in detail.

5.3.4.1 Frame Builder

The *frame builder* is the central element of the frame builder framework. It connects all compound collectors and contains the *timing control*. Compounds that are sent from above are being multiplexed into a single connection on the bottom of the frame builder. On the receiving side it delegates incoming compounds to the compound collectors according to path they have taken on the sending side. The frame builder assures that incoming compounds follow the same path like on the sending side.

5.3.4.2 Timing Control

The *timing control* resides in the frame builder and keeps the activation sequence of the compound collectors and activates them according to the activation sequence. The sequence must be defined in the configuration of the frame builder framework and differs between BS, RS and MS. It defines at which point in time a compound collector gets active and performs its task, see Section 5.3.4.3. Since the timing control is a part of the frame builder, it can easily connect to the involved compound collectors. As soon as the timing control has been started, it executes the sequence of activation without any further notice and loops periodically according to the frame duration.

5.3.4.3 Compound Collector

The *compound collector* acts like a gate for the frame configuration framework and prepares compounds that are intended to be transmitted via the frame builder. The compound collector interface offers two kinds of activation. With the *startCollection* command, the compound collector is ordered to collect compounds from FUs above the compound collector. The radio resource scheduler uses this activation to create a schedule of the compound transmissions that will be processed. If the compound collector can operate without collecting compounds, it may skip this command. With the *start* command the compound collector is ordered to send the compound collection. If the compound collector is connected to the frame builder directly the compounds go straight to the frame builder. The framework allows to use immediate FUs between the compound collector and the frame builder. Figure 5.13 shows three compound collectors in the frame configuration framework. Two of them are directly connected to the frame builder. The third compound collector connects to a FU that is connected to the frame builder.

5.3.5 Component Services

Services are essential in the modular software design to establish communication across layer boundaries. Components provide services to either control the functions of the respective layer or to provide user data transmission services. Services are distinguished by their access control rights.

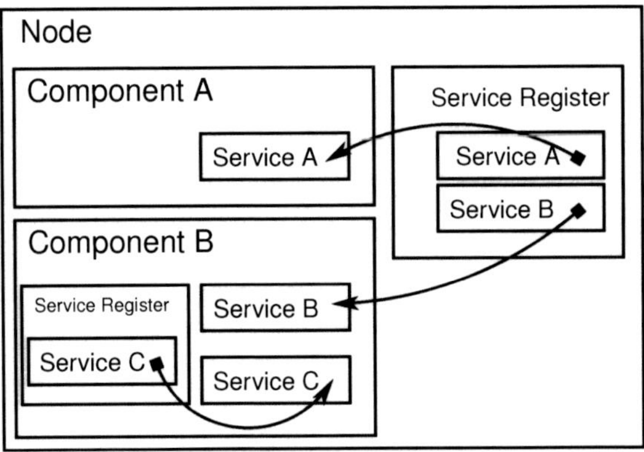

Figure 5.14: Service Register

5.3.5.1 Cross-layer Protocol Services

Cross-layer protocol services are accessible from *outside* of the layer. Components register the named service at the service register of the node. Other layers access the service by a named service request call at the node. The node assures that the layers do not register multiple services with the same name. Named cross-layer services are unique for a node. Typical services are service flow establishment, data transmission services and data handler services.

Figure 5.14 shows a node that contains two components, A and B. Both of the components offer services, service A and B. They register the services at the service register of the node with a unique name which makes the services available for other components. The WiMAC registers the data transmission service at the service register that allows to send layer-3 packets via the DLL WiMAC component. It is equivalent to the SAP known from the ISO/OSI reference model.

5.3.5.2 Internal Component Services

Internal layer services are not accessible from other protocol layers or components. Like the node, the component contains a named service register. FUs access the internal services by a named service request call at the component service register. Typical internal component services are classification of data packets and connection management functions.

In Fig. 5.14 component B contains a service named Service C. The component registers the service at the service register of the component which makes the service available for other members of component B but not for other components like component A. In this example Service B may make use of Service C since it may look up Service C in the service register of the component. Service A could not make use of Service C since Service C it is not registered at the service register of the node and Service A does not have access to the service register of component B.

The WiMAC component offers the following services accessible from WiMAC FUs only but are not exposed to other layers.

Connection Manager

Connections that are established during network entry are registered at the connection manager in both peers. The connection manager keeps the connection type, i.e. control connection or user data connection, the connection identifier, the QoS class and the connection peers. In the RS, the connection manager keeps the connections for both, super- and subordinate stations. The *classifier* uses the connection manager service to identify the connection for a compound.

Buffer Observer

Sending bandwidth requests to the BS is essential to maintain uplink traffic for almost all service grants, see Section 3.3.4. The *buffer observer* allows the MS to maintain an overview of the FUs that can keep one or more compounds for UL transmission. FUs that may keep a compound register themselves at the buffer observer in the beginning of the simulation. If requested, the buffer observer counts the bits of all compounds in the registered FUs that belong to a specified QoS class and the buffer observer returns the sum of bits of all waiting compounds in the FUN. The bandwidth request generator shown in Section 5.3.6.1 makes use of the buffer observer service to estimate the needed bandwidth for each station.

Quality of Service Provider Registry

Supporting multiple load generator modules requires to offer a generic interface for QoS classification of PDUs. The *quality of service provider registry* offers a registry for QoS provider for all

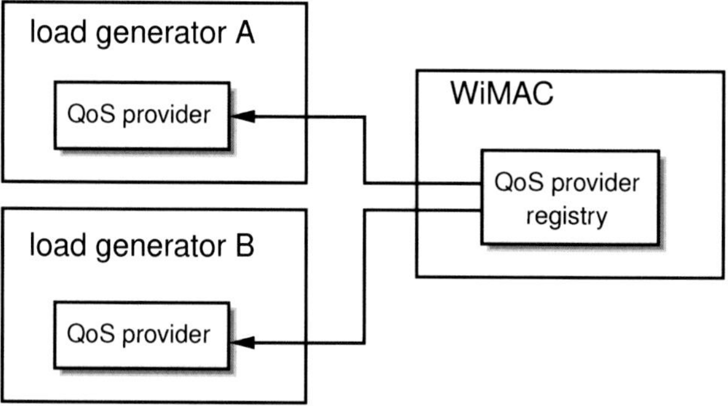

Figure 5.15: QoS provider registry

load generators and gives information about the QoS class of an individual PDU. After all components of the node has been created, the load generator components register their QoS provider at the QoS provider registry. The QoS provider allows the QoS provider registry to identify the QoS class for individual PDUs if the PDU originates from the component of the requested QoS provider. Otherwise the QoS provider returns an error which allows the registry to test the PDU with the next QoS provider until the QoS class of the PDU has been identified.

5.3.6 Stations

The WiMAC module offers three configurations for FUN and protocol services that represent BS, RS and MS respectively. Each station comprises a PHY-layer and the DLL. In addition, the BS and the MS comprises IP, a simplified transport layer and the load generators. In the following, the configuration of the FUN in the DLL is shown.

5.3.6.1 Base Station

The major purpose of the BS DLL instance in the simulator is the radio resource management for associated MS and RS. The BS controls the radio resource allocation for uplink and downlink. For this the BS takes the bandwidth requests of MSs into account and decides which MS to serve.

Figure 5.16 shows the FUN of the BS. The BS maintains a SAP in the upper convergence FU that acts as SAP for higher layers. It receives outgoing packets from higher layers and delivers incoming packets to the respective higher layer. The command of the *upper convergence* contains the source, the target MAC address of the payload data and a protocol type identifier that identifies the protocol layer the payload data originates from.

Incoming compounds delays are measured right after the upper convergence and outgoing compounds are marked by the packet probe. The *packet probe* FU takes measurements on the basis of packets. The probe measures packet delay and packet sizes for both, incoming and outgoing packets. In the terms of the *packet probe* the packet delay means the time span a packet takes to pass a packet probe FUs on the one side and enters a peer *packet probe* on the other side. To measure the time span, the *packet probe* command contains the time the packet passes the FU on the sending side. The packet probe distinguishes between incoming and outgoing packet delays. Incoming packet delays can be easily calculated by subtracting the current time by the stored time in the command. Outgoing packet delays are processed at the sending FU. The receiving *packet probe* notifies the *packet probe* at the sending side about the reception of the packet and the time the packet has taken to arrive at the receiver. The sending *packet probe* processes the measurement as outgoing packet delay.

The *classifier* FU maps a packet to a connection and assigns the CID to the data packet. As the system operates on connections only, the classifier keeps a set of rules that maps packets to connections. The rules depend on the type of the higher layer and the service flow. The *classifier* implements the core function of the IP-CS.

The *multiplexer* and *de-multiplexer* provide multiple ports on the upper or lower side respectively. The purpose of the FUs are to direct compounds according to the QoS attribute along a path. The de-multiplexer guides outgoing compounds according to the de-multiplexer strategy and the multiplexer merges both paths onto a single connection. The multiplexer tags outgoing compounds with a multiplexer opcode which makes it possible for peer multiplexers to direct incoming compounds to the respective path. As a result a peer-multiplexer directs incoming compounds into the same path as they have passed on the sending side. Since different protocols allow to support several type of services, the multiplexer and de-multiplexer can be configured to serve arbitrary number of QoS classes.

Compounds for different QoS classes take different paths in the FUN after the de-multiplexer. Compounds that belong to an interactive service flow, like VoIP traffic are marked by the packet probe in the *interactive* path. The simulator supports PHS with the *compressor* FU that reduces the size of a compound to a predefined value, see Section 4.4.2. The compressor reverts the compression of payload data for incoming packets on the receiving side.

The *buffer* FU stores compounds that cannot be sent immediately. It stores the compounds for further processing until the resource scheduler request more compounds.

To ensure a maximum packet delay, the FUN provides a path for compounds that are about to expire soon. The *deadline gate* allows to pass compounds that are close to end of their time to live. The scheduler assures, that compounds in high urgency are served first.

All paths contain a *marker* FU that tag the chain of activated commands right before the compounds enter the QoS multiplexer. The reflector FUs in the RS use the tag to create a partial copy at the marked position, see also Section 5.3.6.3. As a result, compounds that are forwarded in the RS contain the command data of the marker and above.

The segmentation and reassembly (SAR) FU provides a table of segments in its command to maintain control of segments of a compound. The SAR does not perform segmentation of the compound by itself. The resource scheduler creates segments according to the needs of the scheduling process and keeps an account of the segmentation in the table.

Outgoing compounds pass the hybrid ARQ and error model FUs without changes. For incoming compounds the error model evaluates the packet error probability according to Section 4.3, that is used by the hybrid ARQ FU to initiate a re-transmission at the sending ARQ, if needed.

The signal probe FU forwards outgoing compounds without changes, but takes measurements about the signal strength and interference power for each incoming segment. Right after the signal probe, compounds are directed into the DL scheduler that allocates radio resources for the transmission of the compound segments.

The purpose of the super frame header (SFH) generator is to create SFH compounds whenever the frame builder FU starts the radio super-frame. An additional signal probe takes measurements about signal strength and interference power of the SFH only, which is equivalent to the SINR of the broadcast channel.

The MAP handler FU in the BS creates MAP compounds that indicate the resource allocation for both, DL and UL. The frame builder FU assures that the MAP handler creates the MAP compounds in time.

The UL scheduler in the BS is fed by a *bandwidth request generator* for each QoS class. The bandwidth request generator creates bandwidth request compounds that correspond to the needed UL resources of the MSs. The bandwidth request generator uses the *buffer observer* in the target MS to calculate the bandwidth demand for each station and QoS class, see also Section 5.3.5.2. Bandwidth requests are not sent via the radio channel but are created in the FUN of the BS. The bandwidth request generators are connected via a multiplexer with the UL scheduler that takes the bandwidth requests as input for the scheduling process.

At the bottom of the FUN the frame builder FU connects to the lower convergence that establishes the connection to the RISE component.

5.3.6.2 Mobile Station

Figure 5.17 shows the FUN of the MS that differs from the FUN of the BS occasionally. Unlike the BS, the MS does not contain any bandwidth request generators, since the MS does not respond to bandwidth requests. Instead, the UL scheduler in the MS receives the UL resource allocations from the BS via the MAP handler FU. Based on the MAP information, the UL scheduler triggers UL transmissions on the assigned resource units. In addition the *forwarder* directs incoming compounds to the signal probe without changes and acts as a replacement for the DL scheduler that is not needed in the MS.

5.3.6.3 Relay Station

Since the RS combines functions of the BS and the MS, the FUN of the RS contains considerably more functions than the BS or MS. Figure 5.18 shows the FUN of the RS. Several FUs are duplicated to form two paths, namely one for the BS side that keeps compounds that are about to be transmitted to the BS and to handle DL compounds that arrive from the BS. The other side handles incoming UL compounds from MSs and keeps compounds that the RS sends to MSs. For both sides, the RS maintains separated paths for each QoS class.

As Fig. 5.18 shows, the RS does not provide a SAP on the top of the FUN. Compounds can enter the FUN through the lower convergence only. After they pass the frame multiplexer, UL packets from MSs go through the UL scheduler, DL packets from the BS go through the DL forwarder. Like in the BS or MS, compounds ascend the protocol stack via the signal probe, the error model, the hybrid ARQ and the SAR. The connection manager directs the compounds according to their origin, either to the BS or to the MS side of the FUN. Compounds travel through the packet probes and reach the reflector FUs on the top of the FUN. The reflectors re-inject the compounds into the QoS de-multiplexer according to their next target, BS or MS.

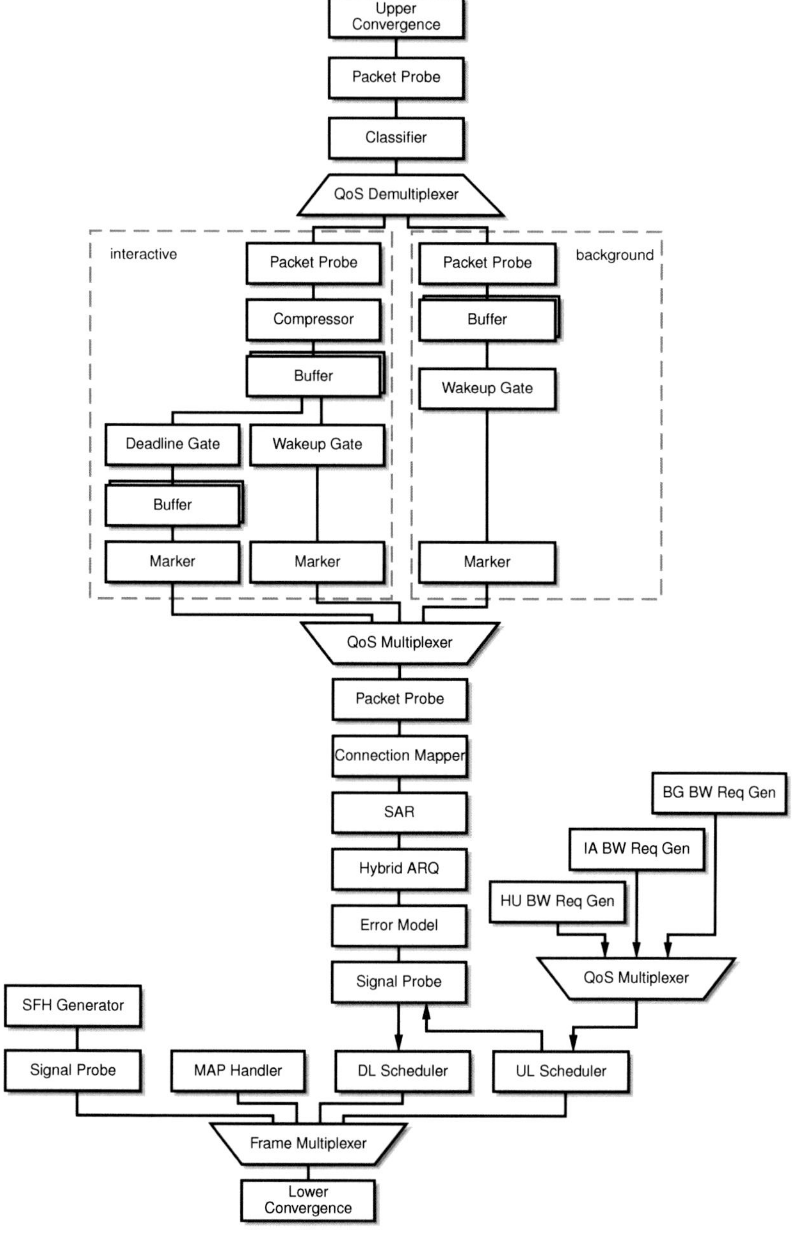

Figure 5.16: Functional Unit Network of the BS node

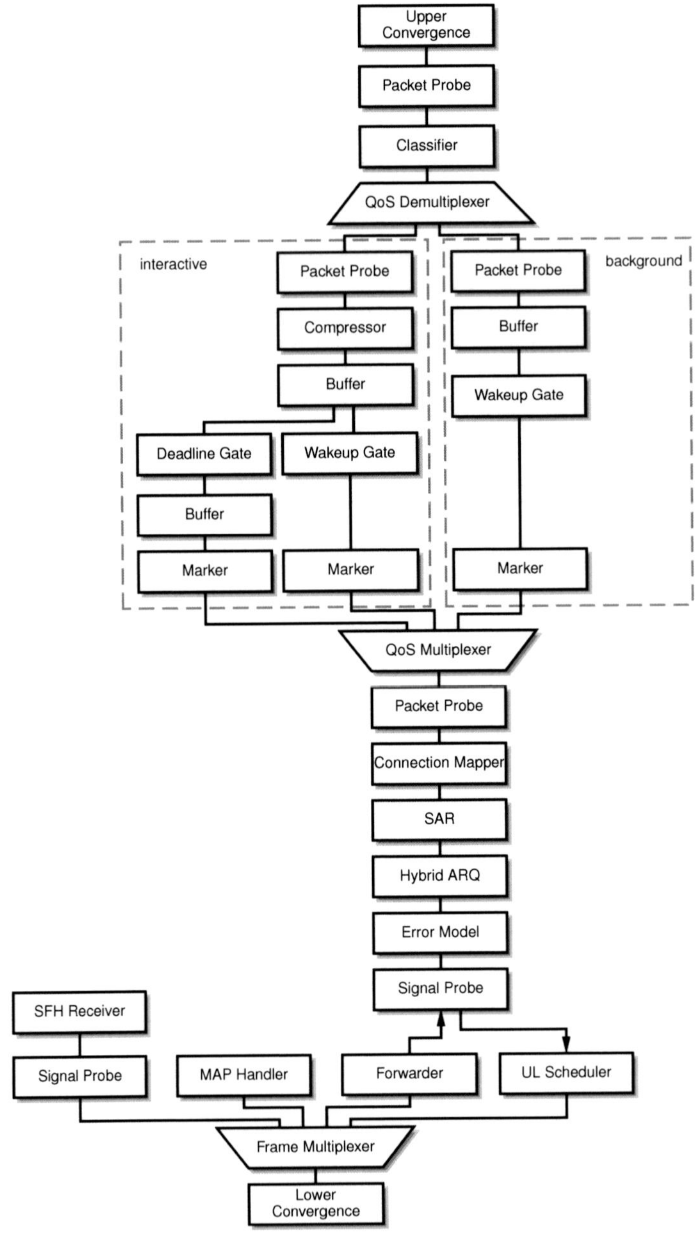

Figure 5.17: Functional Unit Network of the MS node

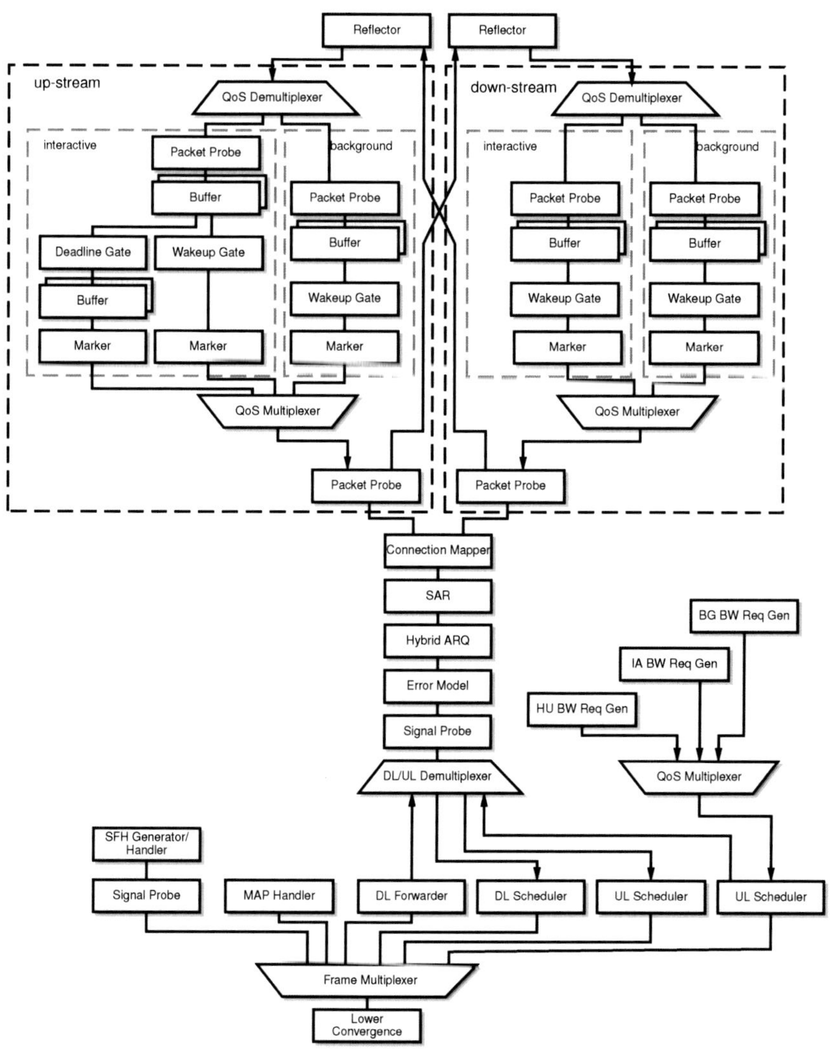

Figure 5.18: Functional Unit Network of the RS node

CHAPTER 6

Performance Evaluation by Simulation

Contents

With the call for proposals of the next generation mobile radio networks of the IMT-A family, the ITU-R released several test environments for candidate systems. This chapter investigates the relay enhanced IEEE-802.16 system in the *base coverage urban* UMa scenario with the help of the openWNS simulator platform and the WiMAC module. The focus of the investigation is on VoIP performance in a multi-cellular network.

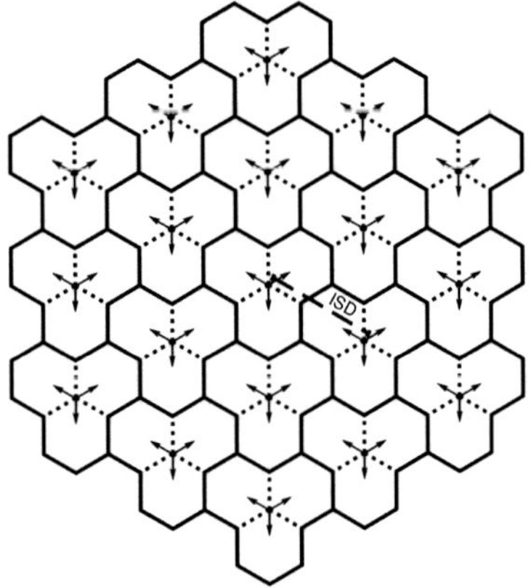

Figure 6.1: Base coverage urban cell scenario

Figure 6.1 shows that the BS site in the test environment mentioned serves three 120 ° cells, each are covered by a sector antenna with antenna characteristics introduced in Section 4.2. Base station sites are arranged according to a regular hexagonal grid with a scenario dependent inter-site distance (ISD). For simulation, 19 sites, each with three cells, are arranged as shown in Fig. 6.1. MSs are distributed uniformly over the whole area. To eliminate border effects, only the center site is evaluated.

The source code of the simulator is available for download under `https://launchpad.net/openwns-allinone`. Configuration files that are used for the following investigation are available online at `http://www.comnets.rwth-aachen.de/~kks/dissertation/index.html`. Simulations can easily be run with the help of the instructions found there.

6.1 Simulation Parameters

To make simulation results comparable across research groups, the IEEE 802.16m task group has published basic simulation assumptions (Srinivasan, 2009) to be followed in system level simulation for performance evaluation, see Table 6.1. The assumptions include antenna configuration, link adaptation and the scenario topology. This work follows the assumptions as far as applicable.

6.1.1 Link Adaptation and Power Control

Table 6.2 gives the parameters b_{BCR} and c_{BCR} required for the BLER calculation under different MCSs, see Section 4.3.2.

The author of (Park et al., 2009) presents a reference SINR to BLER mapping for a wide range of MCSs. Based on that , it is possible to derive a RBIR to BLER mapping that fits into the PHY abstraction model introduced in Section 4.3.2. Table 6.2 show the parameters of the Gaussian cumulative approximation model for all possible MCSs. Figure 6.2 plots the BLER versus mutual information for some example MCSs listed in Table 6.2.

Link adaptation is applied on DL and UL by all stations in the system. A MCS is chosen such that the expected BLER does not exceed 5% for a link with given CSI. While BS and RS use all MCSs shown in Table 6.2, MSs either use QPSK or 16-QAM.

For DL transmissions a fixed spectral power density is used

Table 6.1: Simulation assumptions

Topic	Assumption
DL basic modulation	QPSK, 16-QAM, 64-QAM
UL basic modulation	QPSK, 16-QAM
Duplex scheme	TDD
Multi-antenna transmission	SISO
Data channel coding	convolutional turbo code (CTC)
HARQ	chase combining, asynchronous, maximum 4 transmissions
DL BS transmission power	36 dBm/MHz
DL RS transmission power	11 dBm/MHz
UL transmission power	dynamic power control
frequency reuse	1
bandwidth requests	real-time polling service (rtPS)
simulation time	20 s
ISD	500 m
BS noise figure	5 dB
MS noise figure	7 dB
thermal noise level	174 dB/Hz

as given in Table 6.1. UL transmission power is chosen such that the expected received carrier power does not exceed 15 dB above thermal noise.

6.1.2 Radio Resource Allocation

Radio resource allocation is performed dynamically on a frame-by-frame basis without persistent resource allocations. The resource scheduler allocates radio resources for each MS that are judged as best according to the estimated channel state. The *Kalman filter* is applied for UL. The *Lower Bound* filter is applied for DL, see Section 5.3.3.

Table 6.2: Parameters for Gaussian cumulative approximation

Modulation	Code	Code rate	b_{BCR}	c_{BCR}
QPSK	320	0.12	0.117	0.026
QPSK	720	0.19	0.189	0.033
QPSK	1020	0.27	0.267	0.037
QPSK	1320	0.38	0.379	0.035
QPSK	1520	0.48	0.479	0.027
QPSK	1720	0.61	0.615	0.034
16-QAM	1920	0.39	0.390	0.033
16-QAM	2120	0.48	0.479	0.033
16-QAM	2320	0.61	0.615	0.028
64-QAM	2420	0.46	0.458	0.035
64-QAM	2520	0.52	0.520	0.038
64-QAM	2620	0.61	0.615	0.030
64-QAM	2720	0.69	0.688	0.030
64-QAM	2820	0.78	0.781	0.029
64-QAM	2920	0.82	0.819	0.023
64-QAM	3020	0.92	0.917	0.016

6.1.3 Drop Concept

The simulation of the system performance is carried out as a *drop*, where a drop is defined as one simulation run over the duration of 20 s equivalent of the duration of 4000 radio superframes. To achieve statistical relevant performance parameters the batch means method (Cox & Lewis, 1966) is applied and multiple simulation runs are performed each with different initial seed of the random number generator. Therefore positions of the MSs and radio channel conditions differ in each run. Measurements for a given parameter of all N drops are averaged and presented in the following.

$$\hat{\mu} = \frac{1}{N} \sum_{n=1}^{N} \bar{X}_i \tag{6.1}$$

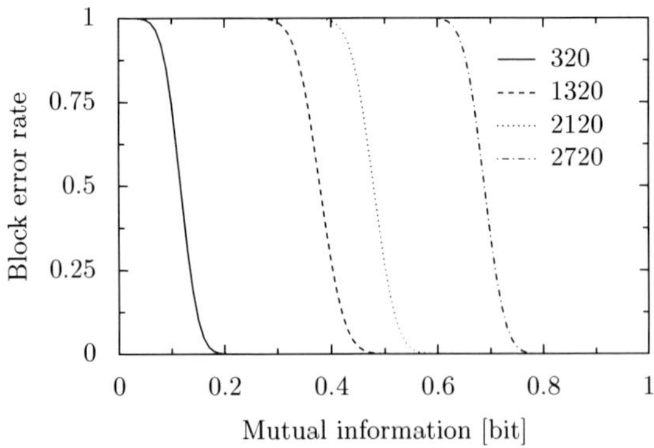

Figure 6.2: Mutual information to BLER mapping for selected codes

$\bar{X}_i$ is the mean of the measured value in drop i. The estimated variance after N drops is calculated according to Eq. (6.2).

$$S^2 = \frac{1}{N-1} \sum_{n=1}^{N} (\bar{X}_i - \hat{\mu})^2 \tag{6.2}$$

The confidence interval is given by Eq. (6.3).

$$\left[\hat{\mu} - z_{(1-\frac{\alpha}{2})} \frac{S}{\sqrt{N}}; \hat{\mu} + z_{(1-\frac{\alpha}{2})} \frac{S}{\sqrt{N}}\right] \tag{6.3}$$

$z_{(1-\frac{\alpha}{2})}$ is the $(1-\frac{\alpha}{2})$-th quantile of the zero-mean normal distribution with variance $\sigma^2 = 1$. For the following a confidence level of $(1-\alpha) = 0.95$ is assumed.

6.2 Frame Configuration

The following investigates the influence of the frame configuration, see Fig. 3.10 on system performance. In addition, three custom configurations shown in Table 6.3 are introduced to improve UL performance for relay enhanced cells.

Table 6.3: Applied frame configuration for RS operation

FCI	duplex mode	access zone DL	access zone UL	relay zone DL	relay zone UL
2	TDD	2	2	2	2
51	TDD	1	1	2	4
53	TDD	1	2	2	3
100	TDD	4	4	0	0

For all figures, a reference scenario without RSs is investigated. For the sake of clarity only results for frame configurations 2, 51, 53 and 100 are shown where frame configuration index (FCI) 100 represents the single-hop scenario without RSs.

6.2.1 VoIP Throughput

Since delay sensitive services are the most critical load to a relay enhanced system owing to multihop radio transmission, we focus in the following on the VoIP service that has strict delay requirements of 50 ms. Of specific interest is the capacity of a relay-enhanced cellular system under VoIP load.

Figure 6.3a shows the DL VoIP cell throughput versus the number of MSs per cell and per bandwidth unit in a cell, i.e. #MS/MHz. The vertical lines give the FCI specific maximum VoIP load (found in Fig. 6.16) that the system can carry.

The figure shows that the frame configuration has a substantial impact on VoIP capacity. FCI=51 achieves the highest DL capacity. Figure 6.3b shows the cell throughput in UL direction. Similar to DL, the UL cell throughput depends on the frame configuration. The single-hop VoIP cell throughput (FCI=100) is the lowest on both, DL and UL, namely 23 MSs. With relay-enhanced cells the VoIP capacity with a well-chosen FCI is up to 47 MSs, which is also visible from the satisfied user ratio in Fig. 6.16. RSs with a well-chosen frame configuration are able

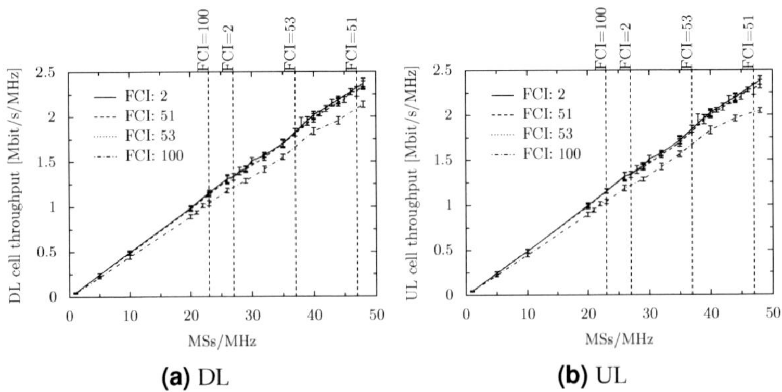

(a) DL **(b)** UL

Figure 6.3: Mean cell throughput for VoIP service

to double the VoIP capacity of the system.

6.2.2 Relay Coverage Area

To evaluate the coverage area of RSs the following compares the amount of MSs that associate to RS and BS. Sine MS are distributed uniformly in the scenario the numbers corresponds to the coverage area.

Figure 6.4a shows the number of associated MSs versus the number of MSs in the cell. It is visible that with relays up to 53 MSs and without relays 49 MSs are associated to the central BS assuming 48 MSs placed in the cell.

Figure 6.4b shows the number of MSs that are associated to a BS. Without relays, the figure corresponds to the results shown in Fig. 6.4a. With RSs, the amount of MSs associated to the BS drops down to 30 %. RSs serve the remaining MSs.

Figure 6.5 shows the radio coverage area by RSs light gray and the coverage by the BS dark gray for the central site (three sectors) as found in (Sambale, 2013). It is clearly visible that RSs cover the outer edge of the cell and even reach into the cell area

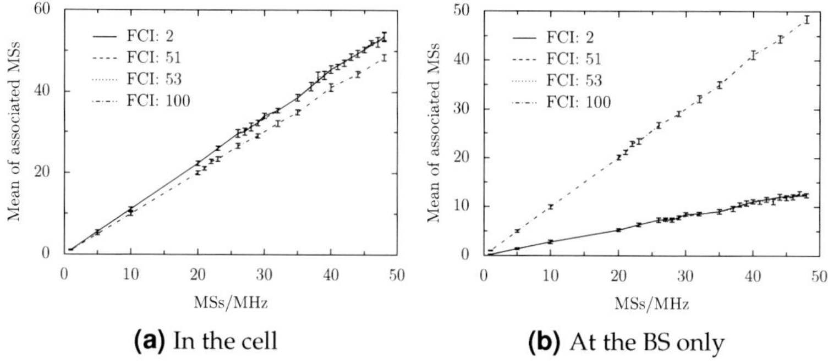

(a) In the cell **(b)** At the BS only

Figure 6.4: Mean number of MSs associated versus load

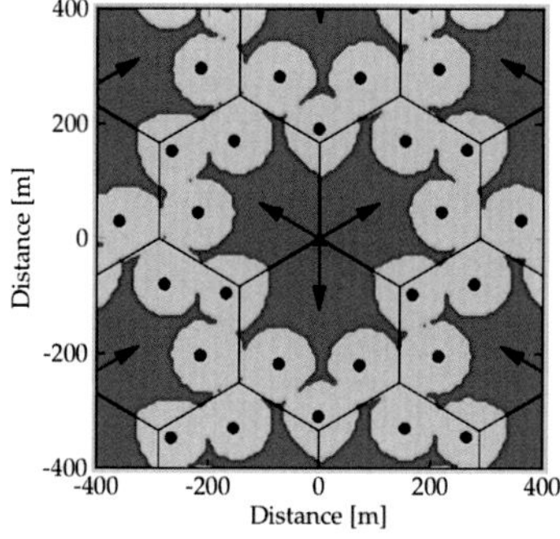

Figure 6.5: Best-server-map with three RSs per cell and association of MSs according to maximum mean SINR (Sambale, 2013, Fig. 5.11)

of the neighbor BSs. Cell edges between sectors of the same site are not served by RSs. It has also been found in (Sambale, 2013, Fig. 5.14) that ca. 65% of the cell area is covered by RSs. This corresponds to the results shown in Fig. 6.4b.

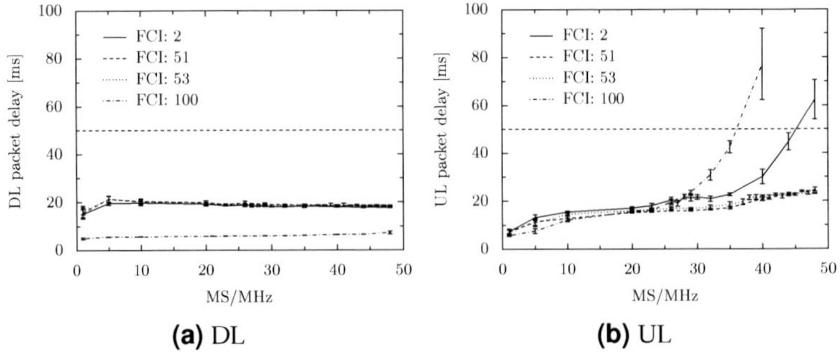

(a) DL **(b)** UL

Figure 6.6: 95-Percentile of VoIP packet delay

6.2.3 VoIP Packet Delay

Figure 6.6a shows the 95-percentile of the end-to-end DL packet delay for VoIP packets. The 50 ms delay limit specified for VoIP traffic is shown. The single-hop scenario (FCI=100) can preserve 95-percentile DL packet delay for up to 48 MS/MHz even below 10 ms. With RSs the 95-percentile of the DL packet delay doubles to 20 ms for up to 48 MS/MHz.

Single hop and multihop UL packet delay for low traffic load are similar, see Fig. 6.6b. For single hop cells the 95-percentile of 50 ms is reached at 36 MS/MHz. Relay enhanced scenarios maintain a packet delay below 30 ms even with 48 MS/MHz where packet delay for FCI=2 is always higher than with FCI=51 or 53. Please note that these figures are of less importance since VoIP capacity has been found in Fig. 6.3 to be 47 MS/MHz at best. Figure 6.6b clearly shows that UL is the bottleneck of the system and RSs improve VoIP packet delay for any configuration.

Figures 6.7a and 6.7b show CDF of packet delay at 25 MS/MHz for DL and UL respectively. In the single-hop scenario (FCI=100) almost all packets are delivered within one radio frame ($\leq$ 5 ms). For multi-hop scenarios packet delay is below 40 ms.

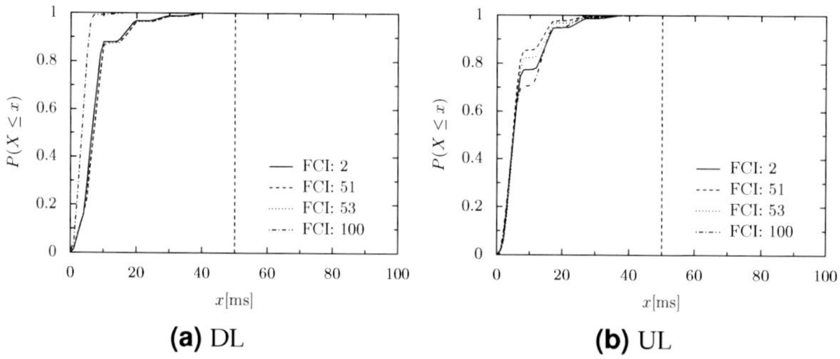

Figure 6.7: CDF of VoIP packet delay with 25 MS/MHz per cell

For UL traffic, the delay CDF of the single-hop scenario appears to be unfavorable owing to more transmit errors compared to multi-hop scenarios but VoIP packet delay is always below 40 ms. Figure 6.7b shows that the bottleneck of the system is found in UL. The system fails to deliver VoIP packets in time for UL while DL packet delay show satisfactory results.

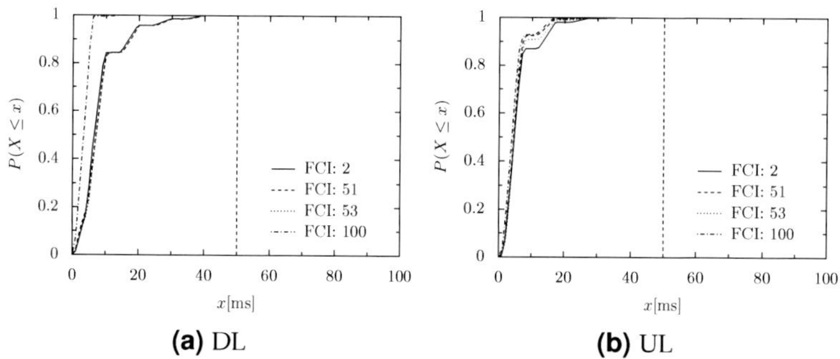

Figure 6.8: CDF of VoIP packet delay with 10 MS/MHz per cell

As Figs. 6.8a and 6.8b show that even with low load of 10 MS/MHz packet delay for FCI 2 and FCI 51/53 is different.

UL packet delay improves more than DL delay under lower load.

6.2.4 SINR and Channel State Estimation Error

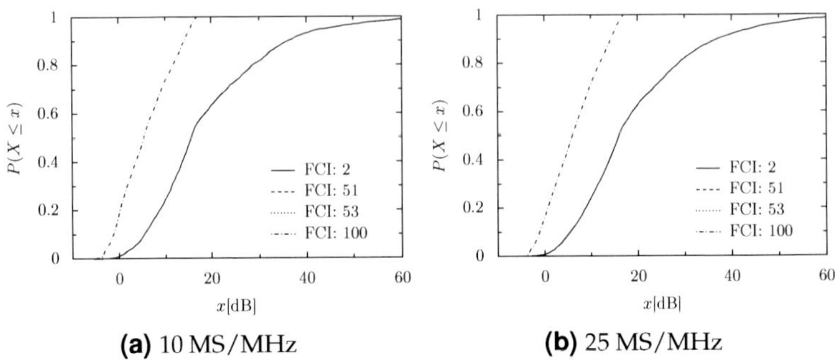

(a) 10 MS/MHz **(b)** 25 MS/MHz

Figure 6.9: SINR CDF of broadcast control channel

Figures 6.9a and 6.9b show the SINR CDF measured by MSs in the broadcast control channel. Since SINR for relay enhanced scenarios only on distances of the MS to BS or RS, and to interfering neighbor BSs or RSs, the CDF does not depend on the frame configurations denoted by FCI and is the same for FCI 2, 51 and 53. It is visible that cells with RSs improve SINR significantly for both, 10 and 25 MS/MHz traffic load.

To gain a deeper insight, how RSs improve the SINR of the broadcast control channel, Figs. 6.10a and 6.10b show the SINR CDF separately for MSs associated to RSs and stations associated to the BS. Clearly MSs associated to RSs achieve significant better SINR compared to MSs associated to the BS. There are almost no stations in the cell experiencing SINR below 0 dB.

The SINR CDF of the broadcast control channel indicates that a relay enhanced system has the potential to increase capacity compared to a BS-only system. Figures 6.11a and 6.11b show

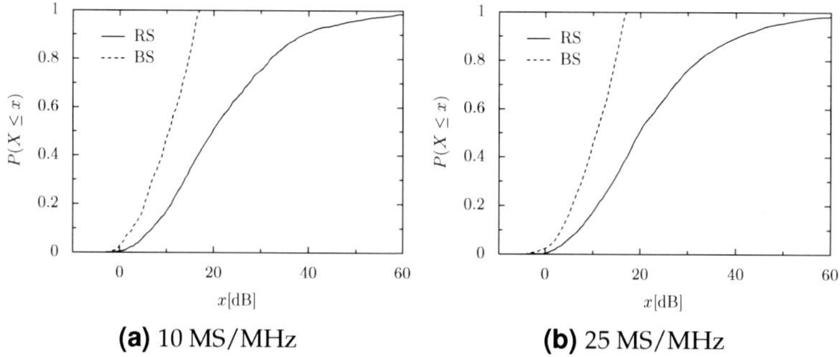

Figure 6.10: SINR CDF of broadcast control channel by station type

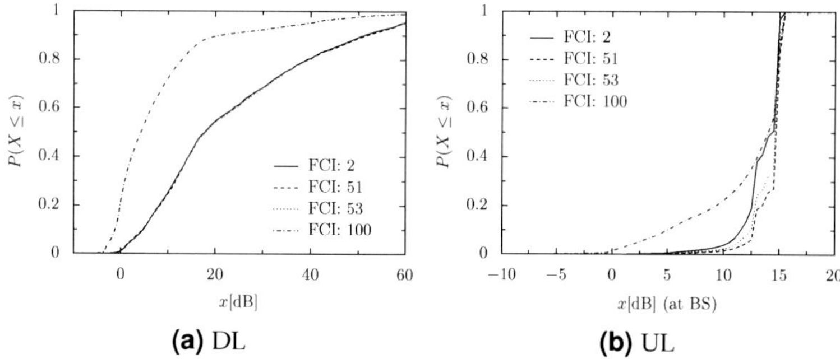

Figure 6.11: SINR CDF of user data frames with 10 MS/MHz

SINR CDF as measured during user data transmission. On DL without relays 90% of the user data experience SINR<20 dB. With relays, the system offers significantly higher SINR, similar to what was found for the broadcast channel. More important, with relays there are almost no user data frames received at SINR<0 dB. Without relays 20% of the user data packets are served with SINR<0 dB.

Figure 6.11b shows that UL data packets in relay based cells mostly experience SINR>10 dB.

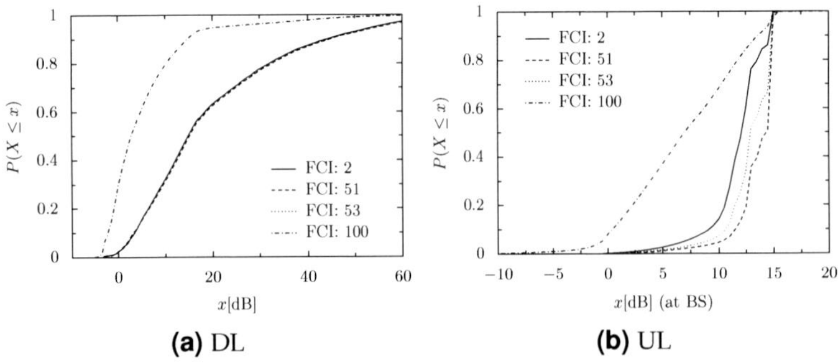

(a) DL **(b)** UL

Figure 6.12: SINR CDF of user data frames with 25 MS/MHz

Figures 6.12a and 6.12b show comparable results for 25 MS/MHz. While the DL SINR CDF is similar to Fig. 6.11a, UL results are different. Without relays MSs achieve significant lower SINR values than visible in Fig. 6.11b, e.g. the 50-percentile reduces from 15 dB to 7 dB. With RSs, the SINR degradation is less obvious. FCI 2 shows the strongest degradation, FCI 51 maintains better results. Nevertheless, configurations that include RSs mostly achieve SINR values above 5 dB.

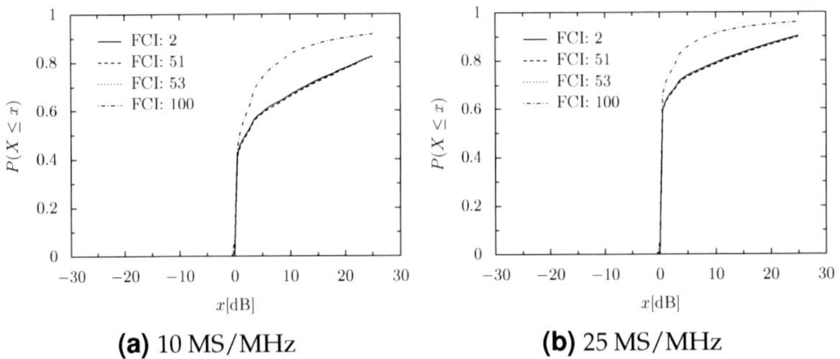

(a) 10 MS/MHz **(b)** 25 MS/MHz

Figure 6.13: CDF of DL SINR estimation error

Channel state estimation errors can either result in bit-errors owing to a wrongly chosen MCS or in waste of radio resources since a too robust message sequence chart (MSC) has been chosen then. Accurate channel prediction is crucial to maximize system performance. Figures 6.13a and 6.13b show DL the CDF of SINR estimation error for 10 and 25 MS/MHz traffic load, respectively. As expected, the DL *worst case* channel estimator applied in the simulator, see Section 5.3.3, never overestimates the current SINR value since it assumes all potential interference sources being active. This results in a one-sided estimation error as shown in Figs. 6.13a and 6.13b. With low VoIP load (10 MS/MHz) 50% of the channel estimation is correct. Without relays the channel prediction seems to be more accurate than with relay-enhanced cells since channel state is better estimated. With high VoIP load channel state prediction appears to be more accurate for both cases, with and without relays. Channel state prediction errors result in selection of more robust MCSs, although more efficient MCSs could be applied. It is expected that channel state estimation is most accurate close to the capacity limit of the system which is about 25 MS/MHz for single hop and 45 MS/MHz for multihop transmission, see Fig. 6.16.

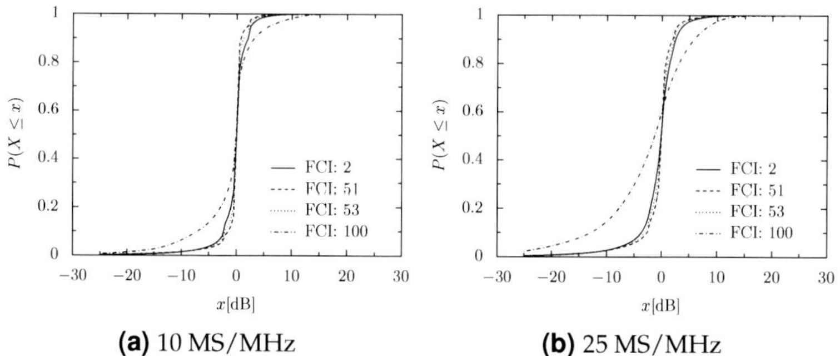

(a) 10 MS/MHz **(b)** 25 MS/MHz

Figure 6.14: CDF of UL SINR estimation error

Channel state prediction by BS/RS for UL transmission is more difficult to acquire since no possibility exists to define a save worst case state. Interference source for UL transmission can be any other MS of neighbor cells. Figure 6.14a shows that channel state estimation error for UL transmission at low VoIP load improves when using relay stations and improves with traffic load, see Fig. 6.14b.

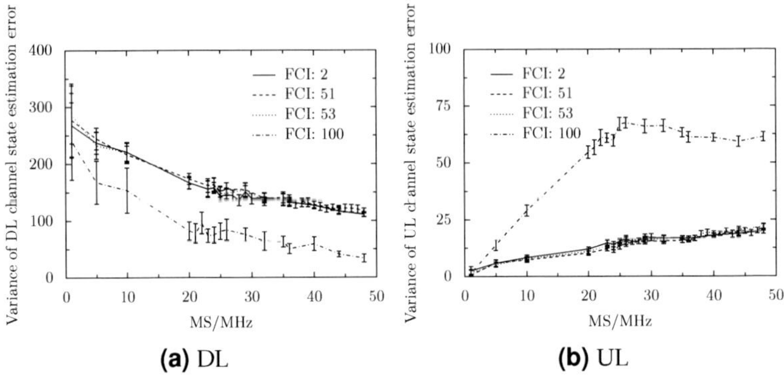

(a) DL

(b) UL

Figure 6.15: Variance of SINR estimation error

Figure 6.15a show that the variance of the channel state estimation error for DL transmissions is high, with low VoIP traffic load in general. The variance decreases with increased VoIP traffic load but is always higher with RSs than without. Figure 6.15b clearly shows that the variance of UL channel state estimation error is substantially reduced if RS are used in a cell. And this is what makes RSs so attractive.

6.2.5 User Satisfaction

User satisfaction as specified in Section 4.4.2 (and specified the same in the IMT-A requirements (ITU-R, 2008b)) is shown in Figs. 6.16a and 6.16b for DL and UL, respectively with confi-

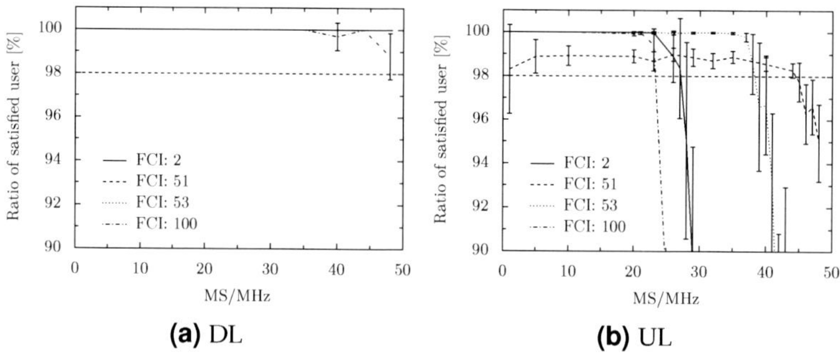

(a) DL **(b)** UL

Figure 6.16: Ratio of satisfied users in percent

dence intervals with 5% level of confidence. On DL the system satisfies VoIP users for the whole range of investigated traffic load without and with RSs. On UL without RSs up to a traffic load of 25 MS/MHz using VoIP service are satisfied. When using RSs, up to 45 MS/MHz can be served with satisfaction. This means the VoIP traffic capacity of relay enhanced systems is twice that of a single hop cellular system.

6.3 VoIP plus Background Traffic

In real world systems a mixture of traffic load has to be carried to fill-up the radio resources that cannot be used by the VoIP service owing to its strict delay requirements. Sine an ongoing radio transmission of a data block cannot be interrupted, mixed traffic load negatively affects VoIP capacity and it is the purpose of the following study to find out the degree of VoIP capacity degradation to be expected under mixed traffic load. The same scenario as before is assumed but with an additional background (BG) Poisson-traffic of 5, 10 and 25 kbit/s for both, UL and DL. The VoIP traffic is assumed prioritized over background traffic. To limit the number of result curves, only the scenarios without relay and with relay under FCI=51 is shown.

6.3.1 Throughput under Mixed Traffic

Since the scheduler prioritizes VoIP traffic it is expected that the results are more or less unaffected by background traffic load.

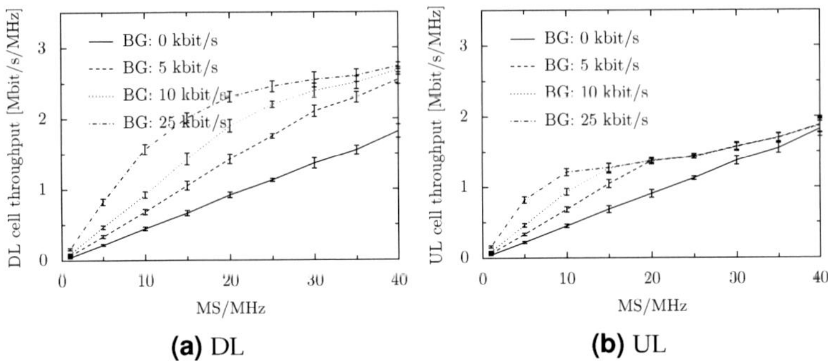

(a) DL **(b)** UL

Figure 6.17: Mean cell throughput without RSs

Figures 6.17a and 6.17b show the total cell throughput (including VoIP traffic) for UL and DL without RSs with a BG traffic load of 0, 5, 10 and 25 kbit/s. It is visible that cell throughput substantially grows steeply with increased background load reaching throughput capacity 2.5 Mbit/sMHz on DL and 2 Mbit/sMHz on UL.

Cell throughput with RSs is lower on DL, see Fig. 6.18a. The reason for that is the variance in the number of MSs associated to a RS, causing overload at some RSs and underload to others. Overloaded RSs drop BG traffic packets to the advantage of VoIP calls, not visible here. Accordingly cell throughput reduces under high BG traffic load below that of lower BG traffic load.

UL cell throughput, see Fig. 6.18b, corresponds to that of RS-less cells, see Fig. 6.17b. RSs improve UL throughput under higher number of MS/MHz.

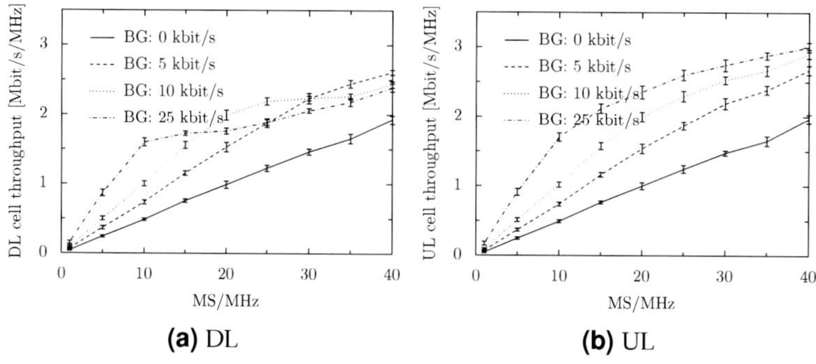

(a) DL **(b)** UL

Figure 6.18: Mean cell throughput with RSs (FCI 51)

6.3.2 VoIP Packet Delay

Like in Figs. 6.6a and 6.6b the following figures show the 95-percentile of VoIP packet delay for both, DL and UL.

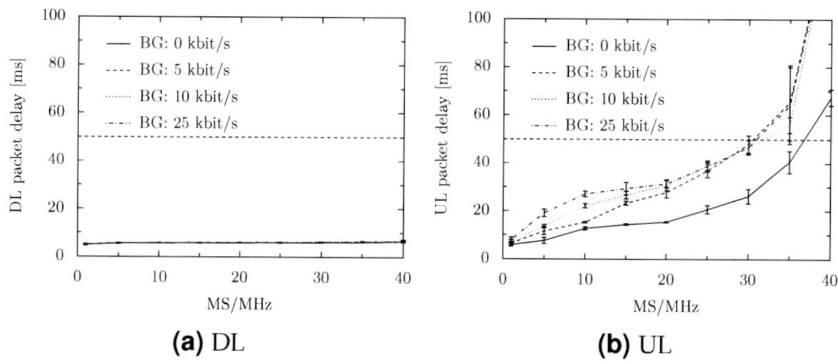

(a) DL **(b)** UL

Figure 6.19: 95-Percentile of VoIP packet delay without RSs

As Fig. 6.19a shows, the 95-percentile of DL packet delay stays at 6 ms independent of the number of MS/MHz. In Fig. 6.19b, the UL shows significantly higher packet delay increasing with the number of MSs/MHz and increasing as BG traffic increases.

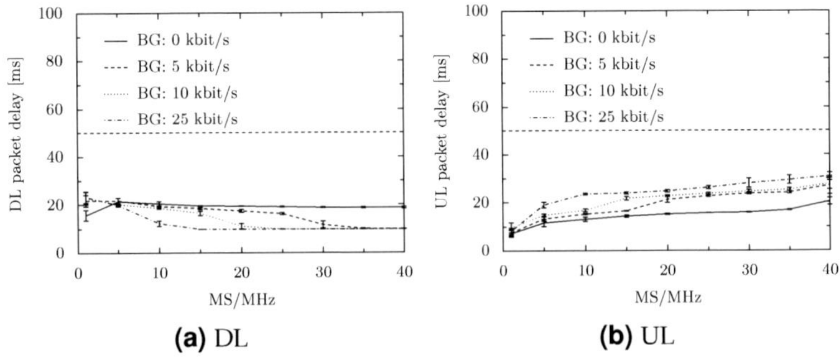

(a) DL **(b)** UL

Figure 6.20: 95-Percentile of VoIP packet delay with RSs (FCI 51)

With relays, DL packet delay is significantly higher. Fig. 6.20a shows the 95-percentile of the packet delay comes close to 20 ms. Different from single-hop scenario, the UL packet delay stays below 30 ms, as Fig. 6.20b shows.

6.3.3 Channel Estimation Error

Background traffic creates a continuous activity on the air interface that may improve channel estimation precision for UL traffic since the variance of the interference level decreases. The probability to take a measurement of the channel when no transmission takes place is reduced. The following figures show the channel estimation error CDF for several background traffic loads.

Similar to the results shown in Section 6.2.4, the channel estimation error for DL traffic is always positive in Figs. 6.21a and 6.21b meaning the channel on DL is never worse than estimated. Increasing the background traffic reduces the channel estimator error variance significantly. With 25 MS/MHz the channel estimation error CDF shows an almost perfect step at 0 dB with heavy background traffic in Fig. 6.21b.

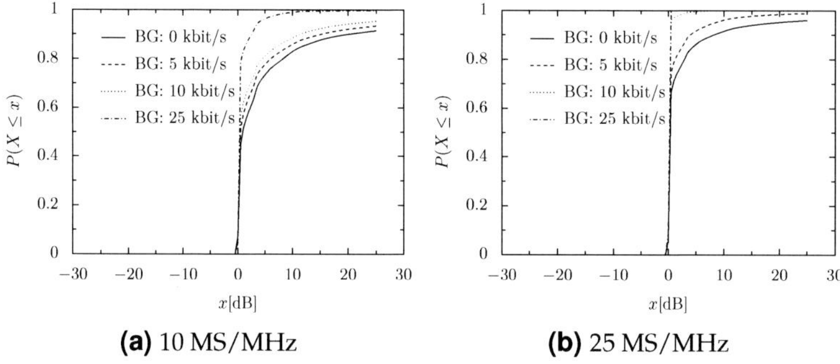

(a) 10 MS/MHz

(b) 25 MS/MHz

Figure 6.21: CDF of DL SINR estimation error without RSs

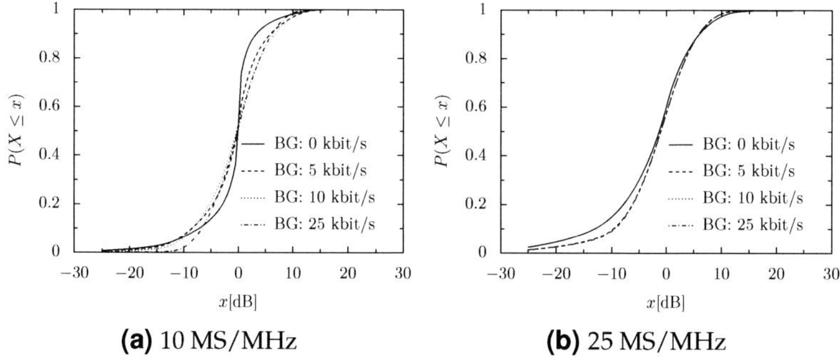

(a) 10 MS/MHz

(b) 25 MS/MHz

Figure 6.22: CDF of UL SINR estimation error without RSs

As Fig. 6.22 shows, background traffic affects the UL channel estimation error much. With only a few number of MS/MHz, background traffic increases the amount of minor channel estimation errors but reduces the amount of large errors. With 25 kbit/s BG traffic there are almost no estimation errors that go below -10 dB. With a larger number of MS/MHz, the effect is less apparent. These results show that background traffic does not improve the UL channel estimation error different from the DL channel estimation error.

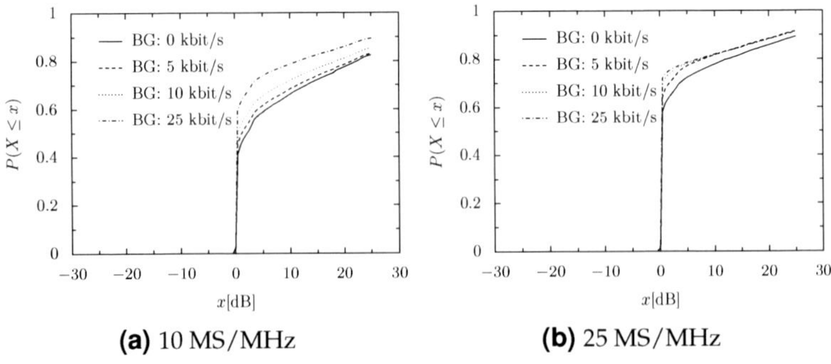

Figure 6.23: CDF of DL SINR estimation error with RSs

RSs reduce the channel estimation error for DL traffic, compare Fig. 6.23 to Fig. 6.21. The channel estimation is less accurate but only drifts towards positive values. Adding background traffic improves the channel estimation precision, but does not affect it significantly.

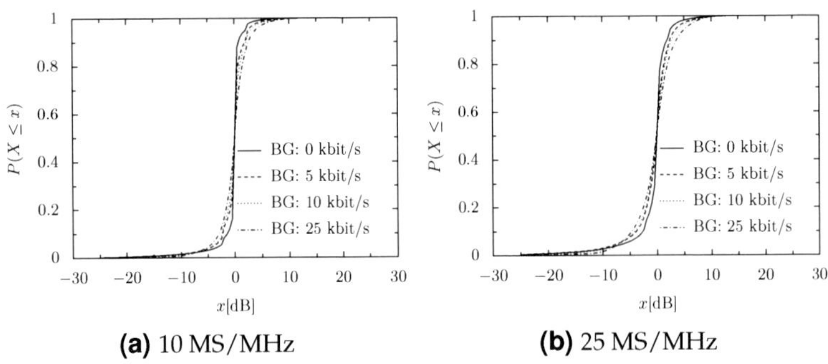

Figure 6.24: CDF of UL SINR estimation error with RSs

Compared to Fig. 6.22 the channel estimation error for UL direction is improved significantly with RSs, see Fig. 6.24. Background traffic does not change the channel estimation error significantly.

6.3.4 User Satisfaction

The protocol should assure a sufficient quality of service for the interactive VoIP users. Figure 6.25a shows that the user satisfaction is met for DL. For UL user satisfaction can hardly be achieved without RSs when background traffic is increasing beyond 5 kbit/s, see Fig. 6.25b.

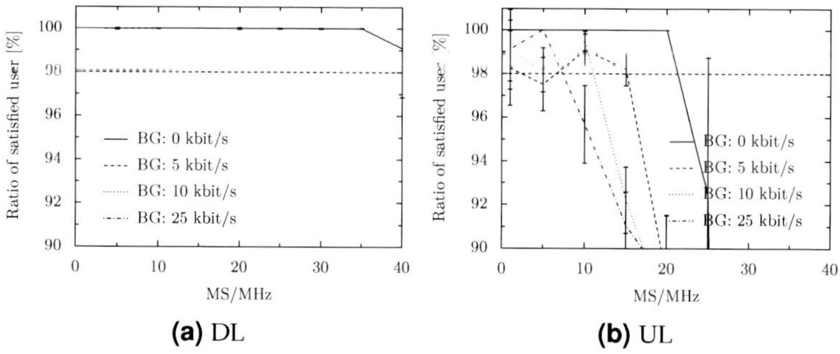

(a) DL **(b)** UL

Figure 6.25: Ratio of satisfied user without RSs

With the help of RSs, the number of satisfied users in DL direction remains unchanged at 100%. For UL, RSs offer a significant improvement, as Fig. 6.26b shows. Although RSs reduce VoIP packet delay, as seen in Fig. 6.20b BG traffic prevents the system to sucessfully serve VoIP users in UL.

Figures 6.27a and 6.27b show the mean VoIP packet loss ratio for DL and UL. For DL the system does not loose any packets independent of the BG traffic load considered. On UL increasing BG traffic causes significantly increasing packet loss due to a bottleneck in radio resources on the access link. Accordingly, with increased BG load. more MSs fail to transmit VoIP packets in time.

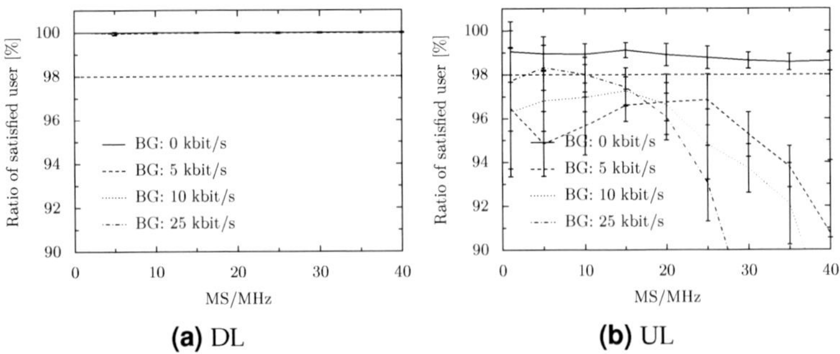

(a) DL **(b)** UL

Figure 6.26: Ratio of satisfied user with RSs (FCI=51)

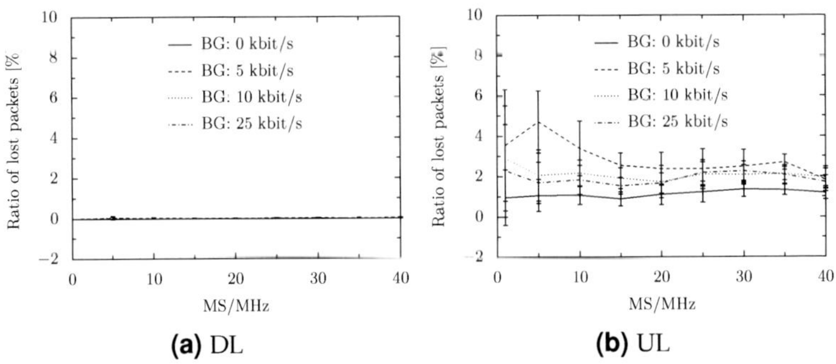

(a) DL **(b)** UL

Figure 6.27: Ratio of lost VoIP packets with RSs (FCI=51)

When comparing Figs. 6.25 and 6.26 it is obvious that BG traffic in the amount considered on DL does not negatively affect VoIP capacity. The UL VoIP capacity with FCI=51 appears to be very sensitive to BG traffic load. Therefore, under BG traffic the radio frame configuration should be dynamically controlled and an appropriate FCI should be chosen dependent on the current BG traffic load to approximately achieve the same VoIP capacity on both, DL and UL. This can be done by switching between the FCIs available from the standard.

6.4 Impact of RS Location on UL Performance Results

In (Sambale, 2013) it is shown that choosing the location of RSs properly in the cell is essential to maximize DL cell capacity. This section investigates the impact of the choice of RS location on UL cell capacity. Starting from the RS locations found in (Sambale, 2013) system performance when shifting RSs locations laterally left and right to an angle α w.r.t. antenna bore sight direction and shifting the centrally located RS from and towards the BS are investigated separately.

6.4.1 Location of laterally placed RS

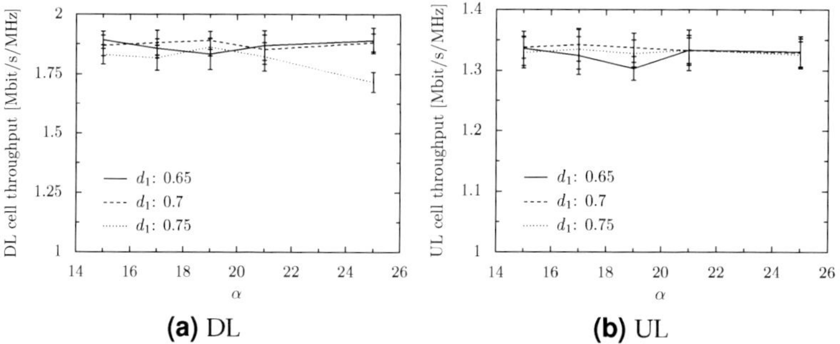

(a) DL **(b)** UL

Figure 6.28: Maximum cell throughput

Figures 6.28a and 6.28b show the cell throughput versus angle α for DL and UL respectively. Results are shown for distances $d_1 = D_1/R$ between the BS and the laterally aligned RSs of 65%, 70% and 75% of the cell radius R, see Fig. 2.9 whilst the location of the RS in antenna boresight direction is kept fixed. The angle between the antenna bore sight and the RS location α varies between 15 ° and 25 °. The results show that the location of the laterally aligned RSs for d_1 in between $0.65R$ and $0.7R$ has only a marginal impact on cell capacity for both, DL and UL.

6.4.2 Location of RS in Antenna Boresight

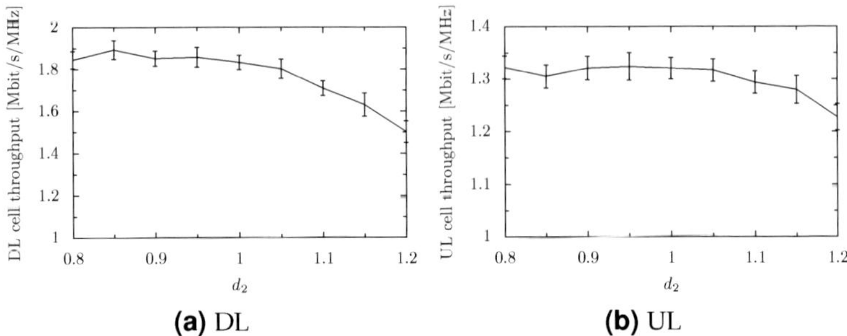

(a) DL **(b)** UL

Figure 6.29: Maximum cell throughput

Figures 6.29a and 6.29b show the DL and UL cell throughput dependent on the distance of the cell edge RS $d_2 = D_2/R$, see Fig. 2.9. Laterally aligned RSs are located at a distance of $d_1 = 0.7$ and $\alpha = 17\,°$. The results show that the DL and UL throughput reaches a maximum at ranges of $0.8 \leq d_2 \leq 1.0$ and decrease with increased distance $d_2 > 1$. Within the statistically uncertainty, DL and UL throughput maximize for the same RS locations.

To achieve maximum cell capacity RSs have to maintain a link to the BS with high SINR and cover the cell area with high SINR that without RSs would suffer from low SINR. These constraints are somewhat in conflict for any given location when considering both, DL and UL.

6.5 Impact of System Bandwidth on Performance Results

As shown in Table 3.6 the system allows bandwidth configurations of 5, 7, 8.75, 10 and 20 MHz. The results presented so far are valid for a system with 2.5 MHz bandwidth which was chosen to reduce computational load when investigating the system by simulation. In the following it is shown for user satisfaction and packet delay how the results presented so far are affected by a larger system bandwidth of 5 and 10 MHz.

Owing to the large memory consumption and simulation time when studying 5 and 10 MHz systems , the number of MS/MHz is limited to a maximum of 25 MS/MHz and FCI=2 is assumed in this study.

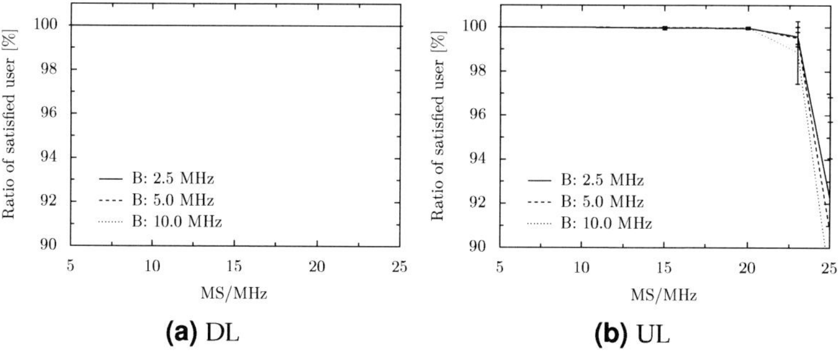

(a) DL **(b)** UL

Figure 6.30: Ratio of satisfied users for single-hop system

As Fig. 6.30a shows, user satisfaction for DL does not change with increased number of MS/MHz for the three system bandwidths considered. Since the system is operated under low load the significance of the results is limited. User satisfaction for UL shows marginal differences for different system bandwidth settings.

As Fig. 6.31a shows, the 95-percentile of the DL packet de-

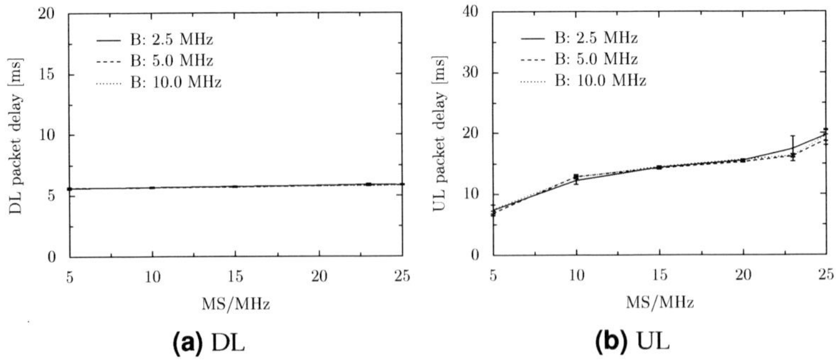

(a) DL **(b)** UL

Figure 6.31: 95-Percentile of packet delay

lay is the same for the three investigated bandwidths. For UL, Fig. 6.31b shows the packet delay differs slightly with increased number of MS. It is therefore concluded that performance results gained for a 2.5 MHz system in tendency are also valid for systems with 5 and 10 MHz bandwidth.

6.6 Validation of openWNS Simulation Results

6.6.1 Validation of Pathloss and SINR CDFs for a System without Relays

With the evaluation of IMT-A candidate systems, the evaluation groups agreed to perform a calibration of the involved system simulators. Pathloss and SINR distributions for each scenario should be compared to guarantee a common basis for the system level performance evaluation. An attempt was made to calibrate the system level simulator openWNS used in this thesis by comparison of its pathloss and SINR CDF to the respective results gained by some organizations participating in the European Celtic project WINNER+ (Mohr et al., 2010).

In the openWNS simulator the UMa scenario has been an-

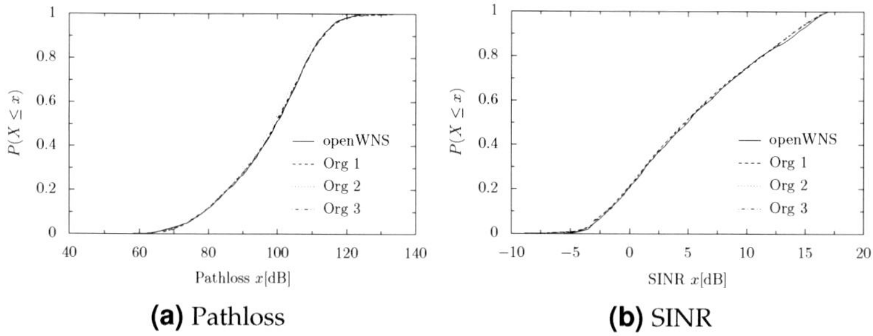

(a) Pathloss **(b)** SINR

Figure 6.32: Reference calibration results from WINNER+ project for UMa scenario

alyzed in several runs to get samples for the radio coverage and signal quality as seen by MSs.In each run 20 MSs are distributed randomly per cell in the investigated simulation scenario, see Fig. 6.1. MSs associate to the best server BS according to the mean SINR derived from measurements of the broadcast channel. Figure 6.32a shows the CDF of the pathloss that MSs achieve with respect to their serving BS in the UMa scenario implemented in the openWNS simulator. In addition results from three WINNER+ organizations published by the WINNER+ project are shown for comparison. The figure shows that the pathloss observed in the openWNS closely fits the reference results.

Figure 6.32b shows the CDF of SINR of the broadcast control channel as gained from the openWNS simulator. Like the pathloss the SINR shows a close fit to the reference results published by the WINNER+ project. Consequently the channel model implementation of the openWNS appears to be calibrated to the other simulators mentioned and can be considered validated by independent simulation results.

6.6.2 System Capacity Validation

System capacity is the maximum system throughput experienced by MSs randomly placed in the UMa scenario. For the validation experiment 20 MSs having full-buffer traffic load are placed in each cell, each MS contributing the maximum possible load to the system for both, UL and DL, see Section 4.4.3. The system schedules MSs to be served rate-fair aiming to serve each MS with the same constant data rate.

To be able to validate the openWNS WiMAC protocol implementation by comparing capacity results to that of an independent second source published work has been searched to find results of a system level simulator having similar detail of protocol and PHY layer implementation. The simulation results presented in (Bou Saleh et al., 2011) gained for an long term evolution advanced (LTE-A) system were found useful for a comparison.

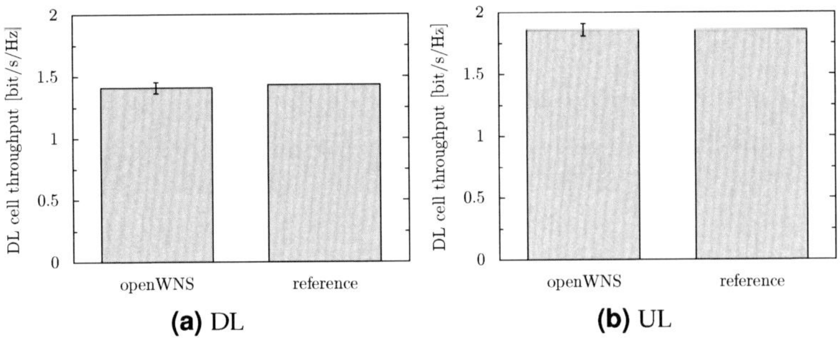

(a) DL **(b)** UL

Figure 6.33: Throughput capacity on MAC layer for systems without relays

From the throughput distribution of MSs shown in (Bou Saleh et al., 2011, Figure 4) the cell capacity in the UMa scenario has been computed. Figure 6.33 compares the cell capacity in both directions, DL and UL, found from (Bou Saleh et al., 2011) to the capacity gained by the openWNS simulator for the BS-only scenario.

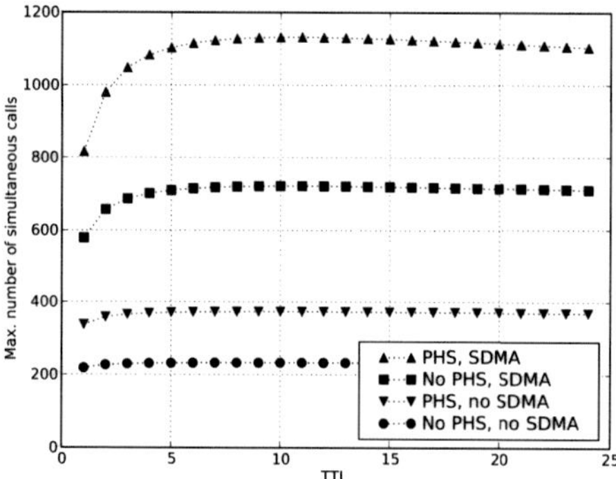

Figure 6.34: Number of served calls versus TTL (Sambale & Klagges, 2009, Fig. 11)

In (Sambale & Klagges, 2009) the number of served VoIP calls in a IEEE 802.16m system with persistent resource allocation has been investigated. Figure 6.34 shows the number of served calls versus the validity period time-to-live (TTL) of a resource allocation in number of radio frames. In (Sambale & Klagges, 2009) an analytic approach has been chosen. To validate the results in this thesis, the results for TTL $= 1$ are of importance. With PHS and without space division multiple access (SDMA) the number of served calls is 360 for 20 MHz namely 18 MS/MHz. This corresponds to 22 MS/MHz found on DL for a system without relays in this thesis.

CHAPTER 7

Conclusions

By implementing the IEEE 802.16m protocol stack in openWNS it is demonstrated that software engineering based on FUs is feasible. FU based protocol software development eases software testing and supports consistence and correctness control from the software engineering point of view compared to con ventional simulator designs. Although runtime performance may be lower ease of protocol configuration from basic protocol elements (FUs), quick modification of the protocol stack and guaranteed software correctness based on unit tests are seen to compensate for runtime disadvantages. Further advantages of FU-based protocol implementation are

- easy insertion of probes for measurement of parameter values
- easy replacement of a FU or FUN and
- usability of protocol emulation software to be ported to a prototype system implementation as known form rapid prototyping.

Simulation experiments performed under conditions as specified by ITU-R with the openWNS based WiMAX protocol have shown that the relay-enhanced IEEE 802.16m system may substantially boost capacity for VoIP service.

The reason for this was found in the increase of SINR on UL and better predictability of the radio channel conditions. Relays were found to reduce percentage of low-performing MSs that in a system without relays decrease overall delay performance of the system substantially.

With three relays per cell overall VoIP capacity grows by up to 50% compared to a system without relays. System capacity is less sensitive to placement of RSs in a cell than expected.

APPENDIX A

SINR to RBIR Mapping

Table A.1 shows the SINR to mutual information mapping that has been applied for the PHY abstraction model of the WiMAC module, see also Section 4.3. The mapping corresponds to the data given in (Srinivasan, 2009) and (Park et al., 2009, Chapter 4).

Table A.1: SINR to RBIR mapping (Srinivasan, 2009)

	QPSK			QAM-16			QAM-64		
SINR Span (dB)	-20:0.5:27			-20:0.5:27			-20:0.5:27		
	0.0072	0.0080	0.0090	0.0036	0.0040	0.0045	0.0024	0.0027	0.0030
	0.0101	0.0114	0.0114	0.0050	0.0057	0.0063	0.0034	0.0038	0.0043
	0.0143	0.0159	0.0159	0.0071	0.0080	0.0089	0.0047	0.0054	0.0060
	0.0200	0.0225	0.0225	0.0100	0.0112	0.0126	0.0067	0.0075	0.0084
	0.0282	0.0315	0.0315	0.0141	0.0158	0.0176	0.0094	0.0106	0.0117
	0.0394	0.0442	0.0442	0.0197	0.0221	0.0247	0.0132	0.0147	0.0165
	0.0551	0.0616	0.0616	0.0276	0.0308	0.0344	0.0184	0.0207	0.0229
	0.0767	0.0855	0.0855	0.0384	0.0428	0.0476	0.0257	0.0285	0.0319
	0.1061	0.1180	0.1180	0.0531	0.0590	0.0656	0.0354	0.0396	0.0437
	0.1456	0.1615	0.1615	0.0728	0.0808	0.0895	0.0488	0.0539	0.0599
	0.1978	0.2184	0.2184	0.0990	0.1094	0.1206	0.0660	0.0732	0.0805
	0.2650	0.2910	0.2910	0.1329	0.1461	0.1603	0.0890	0.0974	0.1073
	0.3489	0.3806	0.3806	0.1756	0.1920	0.2094	0.1172	0.1285	0.1398
	0.4493	0.4859	0.4859	0.2279	0.2474	0.2680	0.1525	0.1653	0.1795
	0.5628	0.6024	0.6024	0.2896	0.3122	0.3357	0.1937	0.2092	0.2247
RBIR Value	0.6817	0.7207	0.7207	0.3600	0.3852	0.4112	0.2415	0.2583	0.2763
	0.7944	0.8281	0.8281	0.4379	0.4653	0.4933	0.2942	0.3132	0.3321
	0.8872	0.9119	0.9119	0.5219	0.5509	0.5804	0.3519	0.3718	0.3924
	0.9507	0.9649	0.9649	0.6103	0.6403	0.6709	0.4131	0.4345	0.4558
	0.9842	0.9901	0.9901	0.7014	0.7317	0.7617	0.4778	0.4997	0.5223
	0.9968	0.9983	0.9983	0.7910	0.8193	0.8463	0.5448	0.5677	0.5907
	0.9997	0.9999	0.9999	0.8716	0.8949	0.9158	0.6141	0.6374	0.6611
	1.0000	1.0000	1.0000	0.9343	0.9501	0.9633	0.6848	0.7087	0.7325
	1.0000	1.0000	1.0000	0.9739	0.9821	0.9883	0.7564	0.7802	0.8036
	1.0000	1.0000	1.0000	0.9927	0.9957	0.9976	0.8269	0.8489	0.8708
	1.0000	1.0000	1.0000	0.9988	0.9994	0.9997	0.8904	0.9100	0.9262
	1.0000	1.0000	1.0000	0.9999	1.0000	1.0000	0.9425	0.9547	0.9668
	1.0000	1.0000	1.0000	1.0000	1.0000	1.0000	0.9732	0.9796	0.9840
	1.0000	1.0000	1.0000	1.0000	1.0000	1.0000	0.9883	0.9910	0.9937
	1.0000	1.0000	1.0000	1.0000	1.0000	1.0000	0.9954	0.9971	0.9983
	1.0000	1.0000	1.0000	1.0000	1.0000	1.0000	0.9995	0.9998	1.0000
	1.0000	1.0000		1.0000	1.0000		1.0000	1.0000	

List of Figures

List of Tables

BIBLIOGRAPHY

3GPP. (2012, September). *TS 26.090 V11.0.0 Mandatory Speech Codec speech processing functions; Adaptive Multi-Rate (AMR) speech codec* (Technical Specification). 650 Route des Lucioles - Sophia Antipolis Valbonne - France: 3rd Generation Partnership Project.

802.16m Task Group, I. (2009). *IEEE 802.16 System Description Document (SDD)* (Tech. Rep.). IEEE. Retrieved from `http://www.wirelessman.org/tgm/index.html`

Andrews, J. G., Ghosh, A., & Muhamed, R. (2007). *Fundamentels of WiMAX, Understanding Broadband Wireless Networking* (T. S. Rapport, Ed.). Prentice Hall.

Antipolis, I. S. (2007-2013). *The NS-3 Network Simulator.* Retrieved from `http://www.nsnam.org/`

Bayan, A., & Wan, T.-C. (2010, June). A scalable QoS scheduling architecture for WiMAX multi-hop relay networks. In *2nd International Conference on Education Technology and Computer (ICETC)* (Vol. 5, p. 326 -331).

Berlemann, L., Pabst, R., Schinnenburg, M., & Walke, B. (2005, April). Reconfigurable Multi-Mode Protocol Reference Model for Optimized Mode Convergence. In *Proceedings of european wireless conference 2005* (Vol. 1, p. 280-286). Nicosia, Cyprus.

Bormann, C., Burmeister, C., Degermark, M., Fukushima, H., Hannu, H., Jonsson, L.-E., ... Zheng, H. (2001, July). *Robust Header Compression (ROHC): Framework and four profiles: RTP, UDP, ESP, and uncompressed* (No. 3095). RFC 3095 (Proposed Standard). IETF. Retrieved from `http://`

`www.ietf.org/rfc/rfc3095.txt` (Updated by RFCs 3759, 4815)

Boudreau, G., Panicker, J., Guo, N., Chang, R., Wang, N., & Vrzic, S. (2009). Interference coordination and cancellation for 4G networks. *Communications Magazine, IEEE, 47*(4), 74-81.

Bou Saleh, A., Bulakci, O., Ren, Z., Redana, S., Raaf, B., & Haemaelaeinen, J. (2011). Resource Sharing in Relay-enhanced 4G Networks. In *Proceedings of 11th European Wireless Conference* (pp. 1–8). Vienna, Austria.

Brady, P. T. (1969, February). A Model for Generating On-Off Speech Patterns in Two-Way Conversation. *The Bell System Technical Journal, 48*, 2445–2472.

Brueninghaus, K., Astely, D., Salzer, T., Visuri, S., Alexiou, A., Karger, S., & Seraji, G.-A. (2005, September). Link performance models for system level simulations of broadband radio access systems. In *16th International Symposium on Personal, Indoor and Mobile Radio Communications (PIMRC)* (Vol. 4, p. 2306 -2311). Berlin, Germany.

Bültmann, D., Muehleisen, M., Klagges, K., & Schinnenburg, M. (2009, May). openWNS - The open Wireless Network Simulator. In *15th European Wireless Conference* (p. 205-210). Aalborg, Denmark: VDE Verlag GmbH.

Chan, G. K. (1992). Effects of sectorization on the spectrum efficiency of cellular radio systems. *Vehicular Technology, IEEE Transactions on, 41*(3), 217-225.

Chang, R. Y., Tao, Z., Zhang, J., & Kuo, C.-C. J. (2013, January). Dynamic fractional frequency reuse (D-FFR) for multicell OFDMA networks using a graph framework. *Wireless Communications and Mobile Computing, 13*(1), 12–27.

Chen, H., Xie, X., & Wu, H. (2009, July). A Queue-Aware Scheduling Algorithm for Multihop Relay Wireless Cellular Networks. In *Mobile wimax symposium mws '09* (pp.

63–68). Napa Valley, California, US.

Cox, D. R., & Lewis, P. A. W. (1966). *The statistical analysis of series of events* (Vol. VIII; Methuen, Ed.). Chapman and Hall.

Einhaus, M. (2009). *Dynamic resource allocation in OFDMA systems* (Doctoral dissertation, RWTH Aachen University, Communication Networks). Retrieved from http://www.comnets.rwth-aachen.de/publications/complete-lists/abstracts/2009/diss-einhaus-2009.html

Esseling, N., Pabst, R., & Walke, B. (2005, April). Delay and Throughput Analysis of a Fixed Relay Concept for Next Generation Wireless Systems. In *Proceedings of 11th European Wireless Conference* (Vol. 1, p. 273-279). Nikosia, Cyprus. Retrieved from http://www.comnets.rwth-aachen.de/typo3conf/ext/cn_download/pi1/passdownload.php?downloaddata=727|1

Esseling, N., Vandra, H., & Walke, B. (2000, September). A Forwarding Concept for HiperLAN/2. In *3rd European Wireless Conference* (Vol. 0, pp. 13–18). Dresden, Germany. Retrieved from http://www.comnets.rwth-aachen.de/publications/complete-lists/abstracts/2000/esvawa-ew2000.html

Esseling, N., Walke, B., & Pabst, R. (2004). Fixed relays for next generation wireless systems: Emerging location aware broadband wireless ad hoc networks. In (p. 71-92). Springer.

Gambiroza, V., Sadeghi, B., & Knightly, E. W. (2004). End-to-end Performance and Fairness in Multihop Wireless Backhaul Networks. In *10th Annual International Conference on Mobile Computing and Networking* (pp. 287–301). New York, NY, USA: ACM.

Genc, V., Murphy, S., & Murphy, J. (2008, April). Performance

Analysis of Transparent Relays in 802.16j MMR Networks. In *Proceedings of 6th Intl. Symposium on Modeling and Optimization in Mobile, Ad Hoc, and Wireless Networks.* Berlin, Germany.

Halpern, S. W. (1983, May). Reuse partitioning in cellular systems. In *Proceedings of 33rd Vehicular Technology Conference* (Vol. 33, p. 322-327). Toronto, Canada.

IEEE. (1986). *ANSI/IEEE Std 1008-1987, IEEE Standard for Software Unit Testing* (Tech. Rep.). 345 East 47th Street, New York, NY 10017, USA: American National Standards Institute, The Institute of Electrical and Electronics Engineers.

IEEE. (2012, April). *802.16.1: WirelessMAN-Advanced Air Interface for Broadband Wireless Access Systems* (Tech. Rep. No. D6). The Institute of Electrical and Electronics Engineers.

ISO. (1994, Nov). *Open System Interconnection (OSI) - Basic Reference Model* (Standard No. ISO/IEC 7489-1:1994(E)). International Standardisation Organisation.

ITU-R. (2008a). *Guidelines for evaluation of radio interface technologies for IMT-Advanced* (Tech. Rep. No. M.2135). International Telecommunication Union, Radiocommunication Sector.

ITU-R. (2008b). *Requirements related to technical performance for IMT-Advanced radio interfaces* (Tech. Rep. No. M.2134). International Telecommunication Union, Radiocommunication Sector.

Kusuda, A., Yamamoto, K., & Yoshida, S. (2005, September). Performance Comparison of Multihop Relaying and Rate Adaption in Cellular Communication Systems. In *16th International Symposium on Personal, Indoor and Mobile Radio Communication* (p. 1621-1625 vol. 3). Berlin, Germany.

Lee, S., Narlikar, G., Pal, M., Wilfong, G., & Zhang, L. (2006, April). Admission control for multihop wireless backhaul networks with QoS support. In *Proceedings of Wireless Communications and Networking Conference, WCNC* (Vol. 1, p. 92 -97). Las Vegas, US.

Leung, K. K. (1999, March). A Kalman-filter method for power control in broadband wireless networks. In *Proceedings of Eighteenth Annual Joint Conference of the IEEE Computer and Communications Societies (INFOCOM'99)* (Vol. 2, pp. 948–956). New York, NY, US.

Lin, Y.-D., & Hsu, Y.-C. (2000, March). Multihop Cellular, A New Architecture for Wireless Communications. In *Infocom 2000, nineteenth annual joint conference of the ieee computer and communications societies* (p. 1273-1282 vol. 3). Tel Aviv, Israel.

MacDonald, V. H. (1979, January). The Cellular Concept. *The Bell System Technical Journal*, *58*(1), 15-41.

Mehlführer, C., Ikuno, J. C., Šimko, M., Schwarz, S., Wrulich, M., & Rupp, M. (2011). The Vienna LTE simulators - Enabling reproducibility in wireless communications research. *EURASIP Journal on Advances in Signal Processing*, 14.

Mohr, W., Cabrejas, J., D'Amico, V., Martín-Sacristán, D., F.Monserrat, J., Saadani, A., … Werner, M. (2010, June). *Final conclusions on end-to-end performance and sensitivity analysis* (Technical Report No. D4.2). Celtic Initiative, WINNER+. Retrieved from http://projects.celtic-initiative.org/winner+/deliverables_winnerplus.html

Necker, M. C. (2008, November/December). Interference Coordination in Cellular OFDMA Networks. *IEEE Network Journal*, *22*(6), 12-19.

Nourizadeh, H., Nourizadeh, S., & Tafazolli, R. (2006, September). Performance Evaluation of Cellular Networks with Mobile and Fixed Relay Station. In *Proceedings of 64th Ve-*

hicular Technology Conference (p. 1-5). Montréal, Canada.

OMNeT++. (2014). *What is OMNeT++.* Retrieved from `http://www.omnetpp.org/home/what-is-omnet`

openWNS. (2004-2013). *The openWNS simulator platform.* `http://www.openwns.org`. Retrieved from `https://launchpad.net/openwns-allinone`

OPNET Technologies, I. (2013). *OPNET Wireless Modeler Suite.* Retrieved from `http://www.opnet.com`

Pabst, R., Walke, B., Schultz, D. C., Herhold, P., Yanikomeroglu, H., Mukherjee, S., … Fettweis, G. P. (2004, September). Relay-Based Deployment Concepts for Wireless and Mobile Broadband Radio. *IEEE Communications Magazine, 42,* 80-89. Retrieved from `http://www.comnets.rwth-aachen.de/publications/complete-lists/abstracts/2004/pawasc-commag2004.html`

Park, J., Lim, J. J., & Kim, T. (2009, September). *Summary of Simulator Calibrations for IMT-Advanced* (Tech. Rep.). IEEE.

Sambale, K. (2013). *Cellular Radio Relay Placement for Optimized Capacity* (Dissertation Thesis, RWTH Aachen University, Communication Networks (ComNets) Research Group). Retrieved from `http://www.comnets.rwth-aachen.de/publications/complete-lists/abstracts/2013/disssambale2013.html`

Sambale, K., & Klagges, K. (2009, May). Increasing The VoIP Capacity of WiMAX Systems Through Persistent Resource Allocation. In *Proceedings of 15th European Wireless Conference* (p. 308-313). Aalborg, Denmark. Retrieved from `http://www.comnets.rwth-aachen.de/publications/complete-lists/abstracts/2009/saklew09.html`

Sambale, K., & Walke, B. (2012a, Sep). Cell Spectral Efficiency Optimization in Relay Enhanced Cells. In *Proceedings of the 23rd IEEE Symposium on Personal, Indoor and Mobile Radio Communications (PIMRC)* (p. 1183-1189). Sydney, Australia.

Sambale, K., & Walke, B. (2012b, Apr). Decode-and-Forward Relay Placement for Maximum Cell Spectral Efficiency. In *Proceedings of the 18th European Wireless Conference.* Poznan, Poland. Retrieved from http://www.comnets.rwth-aachen.de/publications/complete-lists/abstracts/2012/kswew12.html

Schinnenburg, M., Debus, F., Otyakmaz, A., Berlemann, L., & Pabst, R. (2005, Nov). A Framework for Reconfigurable Functions of a Multi-Mode Protocol Layer. In *Proceedings of SDR Forum 2005* (p. 6). Los Angeles, U.S.. Retrieved from http://www.comnets.rwth-aachen.de/publications/complete-lists/abstracts/2005/scdeotbepa-sdr2005.html

Schinnenburg, M., Pabst, R., Klagges, K., & Walke, B. (2007, Sep). A Software Architecture for Modular Implementation of Adaptive Protocol Stacks. In *MMBnet Workshop* (p. 94-103). Hamburg, Germany. Retrieved from http://www.informatik.uni-hamburg.de/bib/medoc/B-281-07.pdf

Schöler, T., & Müller-Schloer, C. (2004, March). Design, implementation and validation of a generic and reconfigurable protocol stack framework for mobile terminals. In *24th International Conference on Distributed Computing Systems Workshops* (pp. 362–367). Tokyo, Japan.

Siebert, M. (2000, Mar). Design of a Generic Protocol Stack for an Adaptive Terminal. In *1st Karlsruhe Workshop on Software Radios* (Vol. 0, p. 31-34). Karlsruhe, Germany.

Siebert, M., & Walke, B. (2001, Jun). Design of Generic and Adaptive Protocol Software (DGAPS). In *Proceedings of 3G*

Wireless and Beyond (Vol. 0). San Francisco, US.

Srinivasan, R. (2009, January). *802.16m Evaluation Methodology Document (EMD)* (Tech. Rep.). Institute of Electrical and Electronics Engineers.

Srinivasan, R., & Hamiti, S. (2010, Dec). *IEEE 802.16m System Description Document (SDD)* (Tech. Rep.). IEEE.

Tao, Z., Li, A., Teo, K. H., & Zhang, J. (2007, November). Frame Structure Design for IEEE 802.16j Mobile Multihop Relay (MMR) Networks. In *Global Telecommunications Conference, GLOBECOM '07* (pp. 4301–4306). Washington, DC, US.

TSG-RAN WG1, E. (2006, May). *R1-061374; Downlink inter-cell interference co-ordination/avoidance - evaluation of frequency reuse* (Tech. Rep.). Shanghai, China: 3rd Generation Partnership Project.

Voudouris, K., Athanasopoulos, N., Georgas, I., Tsiakas, P., Manor, D., Agapiou, G., ... Sheashua, R. (2012, May). Performance verification of a prototype WiMAX relay station. *Science, Measurement Technology, IET*, *6*(3), 176 -180.

Wacker, A., Laiho-Steffens, J., Sipila, K., & Heiska, K. (1999). The impact of the base station sectorisation on WCDMA radio network performance. In *50th Vehicular Technology Conference VTC - Fall* (Vol. 5, p. 2611-2615). Amsterdam, Netherlands.

Walke, B., & Briechle, R. (1985, Nov). A local cellular radio network for digital voice and data transmission at 60GHz. In *Cellular & Mobile Comms. Intern. London* (p. 215-225). Retrieved from `http://www.comnets.rwth-aachen.de/publications/complete-lists/abstracts/1985/wabr-85.html`

WiMAX Forum. (2009, February). *Network Architecture (Stage 2: Architecture Tenets, Reference Model and Reference Points) T32-002-R010v04* (Tech. Rep. No. Release 1.0 Version 4).

WiMAX Forum ®.

Wolz, B., Muehleisen, M., Klagges, K., & Einhaus, M. (2010, April). Region Coordination Across Space Division Multiple Access Enhanced Base Stations in IEEE 802.16m Systems. In *2010 IEEE Wireless Communications and Networking Conference Workshops* (p. 7). Sydney, Australia.

Zheng, H., Wu, M., Choi, Y.-S., Himayat, N., Zhang, J., Zhang, S., & Jalloul, L. (2008). *Link Performance Abstraction for ML Receivers based on RBIR Metrics* (Tech. Rep. No. C802.16m-08/119). IEEE 802.16m Broadband Wireless Access Working Group.

Zimmermann, H. (1980, Apr). OSI Reference Model–The ISO Model of Architecture for Open Systems Interconnection. *IEEE Transactions on Communications*, *28*(4), 425-432.

ACRONYMS

2G	2nd generation
3G	3rd generation
3GPP	3rd Generation Partnership Project
4G	4th generation
AAA	authentication, authorization and accounting
ACID	HARQ channel identifier
ACK	acknowledgment
AGMH	advanced generic MAC header
AMR	adaptive multi-rate
ARQ	automatic repeat request
ASN	access service network
ASN-GW	access service network gateway
ATM	asynchronous transfer mode
BG	background
BLER	block error rate
BR	bandwidth request
BS	base station
BW	bandwidth
CAPEX	capital expenditure
CCDF	complementary cumulative distribution function
CDF	cumulative distribution function
CID	connection identifier
CLRU	contiguous logical resource unit
CPS	common part sublayer
CPU	central processing unit
CRC	cyclic redundancy check

CS	convergence sublayer
CSI	channel state information
CSN	connectivity service network
CTC	convolutional turbo code
DF	decode and forward
DHCP	dynamic host configuration protocol
DL	downlink
DLL	data link layer
DLRU	distributed logical resource unit
EESM	exponential effective SINR mapping
EH	extended header
ertPS	extended real-time polling service
FC	fragmentation control
FCFS	first come first serve
FCI	frame configuration index
FDD	frequency division duplex
FDM	frequency division multiplexing
FEH	fragmentation extended header
FFT	fast fourier transformation
FID	flow identifier
FU	functional unit
FUN	functional unit network
GMSH	grant management subheader
GPS	Global Positioning System
GRE	generic routing encapsulation protocol
GSM	Global System for Mobile Communication
GUI	graphical user interface
HA	home agent
HARQ	hybrid automatic repeat request
HU	high urgency
IA	interactive

IE	information element
IEEE	Institute of Electrical and Electronics Engineers
IMT-A	IMT-Advanced
InH	indoor hotspot
IP	Internet protocol
IP-CS	Internet protocol conversion sublayer
IPv4	Internet protocol version 4
IPv6	Internet protocol version 6
ISD	inter-site distance
ISO	International Standardization Organization
ITU	International Telecommunication Union
ITU-R	International Telecommunication Union, Radiocommunication Sector
L1	layer 1
L2	layer 2
LGPL	GNU Lesser General Public License
LLC	logical link control
LoS	line of sight
LRU	logical resource unit
LTE	long term evolution
LTE-A	long term evolution advanced
MAC	medium access control
MAP	medium access pointers
MCS	modulation and coding schemes
MI	mutual information
MIP	mobile internet protocol
MR-BS	multi-hop relay base station
MS	mobile station
MSC	message sequence chart
NACK	non-acknowledgment

NAP	network access provider
NED	network description
NLoS	non line of sight
nrPS	non-real-time polling service
NSP	network service provider
OFDM	orthogonal frequency division multiplex
OFDMA	orthogonal frequency division multiple access
OSI	Open System Interconnection
PCI	protocol control information
PDF	probability density function
PDU	protocol data unit
PEH	packing extended header
PHS	payload header suppression
PHY	physical
PRU	physical resource unit
PtP	point-to-point
QAM	quadrature amplitude modulation
QoS	quality of service
QPSK	quadrature phase shift keying
RBIR	received bit mutual information rate
REG-REQ	register request
REG-RSP	register response
RISE	Radio Interference Simulation Engine
RLC	radio link control
RMa	rural macro
RNG	ranging
RNG-REQ	ranging request
RNG-RSP	ranging response
ROHC	robust header compression
RRCM	radio resource control and management

RS	relay station
RTG	receive-transmit turnaround gap
RTP	real-time transport protocol
rtPS	real-time polling service
RU	resource unit
RX	receive
SAP	service access point
SAR	segmentation and reassembly
SDM	space division multiplex
SDMA	space division multiple access
SDU	service data unit
SFH	super frame header
SFID	service flow identifier
SID	silence insertion descriptor
SINR	signal to interference plus noise ratio
SISO	single input single output
SN	sequence number
SNMP	simple network management protocol
STID	station identifier
STR	simultaneous transmit and receive
TCP	transmission control protocol
TDD	time division duplex
TDM	time division multiplex
TDMA	time division multiple access
TTG	transmit-receive turnaround gap
TTL	time-to-live
TTR	time-division-transmit and receive
TX	transmit
UDP	user datagram protocol
UGS	unsolicited grant service
UL	uplink
UMa	urban macro
UMi	urban micro

UMTS	Universal Mobile Telecommunication System
VoIP	voice over IP
WG	wakeup gate
WiMAC	WiMAX media access control protocol
WiMAX	Worldwide Interoperability for Microwave Access
WINNER	Wireless World Initiative New Radio

LIST OF SYMBOLS

A_a	azimuth component of antenna characteristic
A_e	elevation component of antenna characteristic
b_{BCR}	Gaussian cumulative approximation parameter b
c	vacuum light propagation velocity
c_{BCR}	Gaussian cumulative approximation parameter c
Δf	subcarrier spacing
F_S	sampling frequency
h_{BS}	height of base station
I	mutual information
I_i	mutual information of block segment i
m_i	size of block segment i
N_0	thermal noise power
N_{FFT}	FFT size
N_{sym}	number of consecutive OFDMA symbols per physical resource unit
P_C	carrier power
P_{In}	interference power caused by station n
P_{sc}	number of consecutive subcarriers per physical resource unit
r_{BLE}	block error rate
T_b	useful symbol duration
T_S	symbol duration

LEBENSLAUF

1977	Geboren in Bielefeld
1987–1996	Friedrich Leopold Woeste Gymnasium Hemer
1996–1997	Grundwehrdienst
1997–2006	Studium an der RWTH Aachen im Fach Physik und Elektrotechnik mit Abschluß Diplom Eletrotechnik
2004–2005	Panasonic Research and Development Center Germany, Langen
2006–2013	Wissenschaflicher Angestellter am Lehrstuhl für Kommunikationsnetze, später Forschungsgruppe Kommunikationsnetze

NACHWORT

Die vorliegende Arbeit entstand während meiner Tätigkeit als wissenschaftlicher Mitarbeiter am Lehrstuhl für Kommunikationsnetze der RWTH Aachen.

Mein Dank gilt insbesondere Herrn Prof. Dr.-Ing. Bernhard Walke für die Anregungen zum Thema der Dissertation, die sehr gute, fortwährende Betreuung und Förderung der Arbeit, sowie die kritische Durchsicht des Textes. Herrn Prof. Dr. - Ing. Timm-Giel danke ich für die freundliche Übernahme des Koreferats.

Besonderen Dank geht an meine Kollegen am Lehrstuhl für Kommunikationsnetze Daniel Bültmann, Maciej Mühleisen, Holger Rosier, Klaus Sambale, Marc Schinnenburg und Benedikt Wolz. Bedanken möchte ich mich auch bei allen Diplomanden, Studienarbeitern und studentischen Hilfskräften, die durch ihre Arbeit wesentlich zum Gelingen beigetragen haben: Philipp Michalschik, Joachim Prick, Ramin Rezai Rad und Adel Zalok.

Schließlich bedanke ich mich ebenso ganz herzlich bei denjenigen, die die Mühe des Korrekturlesens auf sich genommen haben sowie meiner Familie für die Unterstützung und Geduld während der Anfertigung dieser Arbeit.

Aachen, im Februar 2015 Karsten Klagges

Aachener Beiträge zur Mobil- und Telekomunikation

ABMT Band 1
Herrmann, C.
Stochastische Modelle für ATM-Konzepte,
1. Auflage 1995, 138 Seiten;
ISBN 3-86073-380-X

ABMT Band 2
Lawniczak, D. R.
Modellierung und Bewertung der Datenverwaltungskonzep-te in UMTS,
1. Auflage 1995, 230 Seiten;
ISBN 3-86073-381-8

ABMT Band 3
Junius, M.
Leistungsbewertung intelligenter Handover-Verfahren für zellulare Mobilfunksysteme,
1. Auflage 1995, 208 Seiten;
ISBN 3-86073-382-6

ABMT Band 4
Steffan, H.
Stochastische Modelle für den Funkkanal und deren Anwen-dung,
1. Auflage 1996, 164 Seiten;
ISBN 3-86073-383-4

ABMT Band 5
Böhmer, S.
Entwurf eines ATM-basierten Funknetzes und Software-Entwurfsmethodik zur Imple-mentierung,
1. Auflage 1996, 172 Seiten;
ISBN 3-86073-384-2

ABMT Band 6
Guntermann, M.
Universelle Benutzermobilität auf der Basis des Intelligenten Netzes - Entwurf, Bewertung und Implementierung -,
1. Auflage 1996, 164 Seiten;
ISBN 3-86073-385-0

ABMT Band 7
Kleier, S.
Neue Konzepte zur Unter-stützung von Mobilität in Telekommunikationsnetzen,
1. Auflage 1996, 204 Seiten;
ISBN 3-86073-386-9

ABMT Band 8
Decker, P.
Entwurf und Leistungsbewer-tung hybrider Fehlersicher-ungsprotokolle für paketierte Sprach- und Datendienste im GSM-Mobilfunksystem,
1. Auflage 1997, 232 Seiten;ISBN 3-86073-387-7

ABMT Band 9
Hußmann, H.
Algorithmen zur Kapazitäts-optimierung schnurloser Mo-bilfunksysteme nach DECT-Standard,
1. Auflage 1997, 180 Seiten;
ISBN 3-86073-388-5

ABMT Band 10
Plenge, C.
Leistungsbewertung öffent-licher DECT-Systeme,
1. Auflage 1997, 258 Seiten;
ISBN 3-86073-389-3

ABMT Band 11
Kennemann, O.
Lokalisierung von Mobilsta-tionen anhand ihrer Funk-meßdaten,
1. Auflage 1997, 162 Seiten;
ISBN 3-86073-620-5

ABMT Band 12
Wietfeld, C. M.
Mobilfunksysteme für die europäische Verkehrsleittech-nik - Leistungsanalyse des CEN-DSRC-Standards -,
1. Auflage 1997, 210 Seiten;
ISBN 3-86073-621-3

ABMT Band 13
Görg, C.
Verkehrstheoretische Modelle und stochastische Simulationstechniken zur Leistungsanalyse von Kommunikationsnetzen,
1. Auflage 1997, 220 Seiten;
ISBN 3-86073-622-1

ABMT Band 14
Shahbaz, M.
Zufallsgesteuerte Verfahren zur Topologieoptimierung von Telekommunikations-netzen,
1. Auflage 1998, 208 Seiten;
ISBN 3-86073-623-X

ABMT Band 15
Fröhlich, H. M.
Mehrwertdienste intelligenter Netze zur Realisierung der universellen, persönlichen Mobilität,
1. Auflage 1998, 208 Seiten;
ISBN 3-86073-624-8

ABMT Band 16
Geulen, E.
Modelle zur Realisierung offener Dienste in zellularen Mobilfunknetzen nach dem GSM-Standard,
1. Auflage 1998, 224 Seiten;
ISBN 3-86073-625-6

ABMT Band 17
Guntsch, A.
Untersuchungen zur Integration terrestrischer und satellitengestützter Mobilfunksysteme,
1. Auflage 1998, 200 Seiten;
ISBN 3-86073-626-4

ABMT Band 18
Petras, D.
Entwicklung und Leistungsbewertung einer ATM- Funkschnittstelle,
1. Auflage 1999, 240 Seiten;
ISBN 3-86073-627-2

ABMT Band 19
Brasche, G. Ph.
Prototypische Bewertung und Implementierung von neuen Paket-Datendiensten für das GSM-Mobilfunksystem,
1. Auflage 1999, 272 Seiten;
ISBN 3-86073-628-0

ABMT Band 20
Walke, Th.
Markteintritt in lokale Telekommunikationsmärkte Eine Untersuchung aktuellen, potentiellen und substitutiven Wettbewerbs im deutschen Teilnehmeranschlußnetz,
1. Auflage 1999, 424 Seiten;
ISBN 3-86073-629-9

ABMT Band 21
Bjelajac, B.
Modellierung und Leistungsbewertung von mobilen Satellitensystemen mit dynamischer Kanalvergabe,
1. Auflage 1999, 176 Seiten;
ISBN 3-86073-822-4

ABMT Band 22
Walke, Bernhard (Hrsg.).
Friedrich Schreiber: Ausgewahlte Kapitel des wissenschaftlichen Werkes,
1. Auflage 2000, 256 Seiten;
ISBN 3-86073-823-2

ABMT Band 23
Hettich, A. B.
Leistungsbewertung der Standards HIPERLAN/2 und IEEE 802.11 für drahtlose lokale Netze,
1. Auflage 2001, 256 Seiten;
ISBN 3-86073-824-0

ABMT Band 24
Lott, M.
Entwurf eines drahtlosen multihop Ad-hoc-Funknetzes mit Dienstgüteunterstützung
1. Auflage 2001, 232 Seiten;
ISBN 3-89653-853-5

ABMT Band 25
Hartmann, J.
Agentenbasiertes Kommunikationskonzept zur Realisierung von E-Commerce-Diensten in zellularen Mobilfunknetzen – Entwurf, Implementierung und Bewertung
1. Auflage 2002, 186 Seiten
ISBN 3-86073-825-9

ABMT Band 26
Kadelka, A.
Entwurf und Leistungsanalyse des mobilen Internetzugangs über HIPERLAN/2
1. Auflage 2002, 220 Seiten
ISBN 3-86073-826-7

ABMT Band 27
Scheibenbogen, M.
Dynamische Kanalvergabe in zellularen Funksystemen
1. Auflage 2002, 200 Seiten
ISBN 3-86073-827-5

ABMT Band 28
Steppler, M.
Leistungsbewertung von TETRA-Mobilfunksystemen durch Analyse und Emulation ihrer Protokolle
1. Auflage 2002, 400 Seiten
ISBN 3-86073-828-3

ABMT Band 29
Obradovic´, V.
Teletraffic Analysis of Mobile Satellite Systems
1. Auflage 2002. 226 Seiten
ISBN 3-86073-980-8

ABMT Band 30
Habetha, J.
Entwurf eines cluster-basierten ad hoc Funknetzes
1.Auflage 2002 360 Seiten
ISBN 3-86073-981-6

ABMT Band 31
Rapp, J. A.
Ratenanpassung in HIPERLAN/2 mit realistischer Interferenz-Modellierung und detailliertem Protokollstapel
1. Auflage 2002, 208 Seiten
ISBN 3-86073-829-1

Aachener Beiträge zur Mobil- und Telekomunikation

ABMT Band 32
Xu, B.
Self-organizing Wireless Broadband Multihop Networks with QoS Guarantee
1. Auflage 2002, 200 Seiten
ISBN 3-86073-982-4

ABMT Band 33
Herwono, I.
Entwicklung und Leistungsbewertung von mobilen Diensten für Electronic Commerce Anwendungen
1. Auflage 2002, 270 Seiten
ISBN 3-86073-983-2

ABMT Band 34
Peetz, J.
Multihop-Ad-hoc-Kommunikation mit dynamischer Frequenzwahl, Leistungssteuerung und Ratenanpassung für drahtlose Netze im 5 GHz Band
1.Auflage 2003, 285 Seiten
ISBN 3-86073-984-0

ABMT Band 35
Mangold, S.
Analysis of IEEE 802.11e and Application of Game Models for Support of Quality-of-Service in Coexisting Wireless Networks
1. Auflage 2003, 287 Seiten;
ISBN 3-86073-985-9

ABMT Band 36
Speetzen, A.
Methodische Beiträge zur redundanten und dynamischen Planung von Kommunikationsnetzen
1. Auflage 2003, 380 Seiten
ISBN 3-86130-164-4

ABMT Band 37
Krämling, A.
Kanalvergabe und Sendeleistungssteuerung bei HiperLAN/2
1. Auflage 2003, 232 Seiten
ISBN 3-86130-165-2

ABMT Band 38
Stuckmann, P.
Traffic Engineering Concepts for Cellular Packet Radio Networks with Quality of Service Support
1. Auflage 2003, 322 Seiten
ISBN 3-86130-166-0

ABMT Band 39
Vornefeld, U.
Verkehrstheoretische Dimensionierungsverfahren für paketvermittelnde Mobilfunknetze
1. Auflage 2003, 358 Seiten
ISBN 3-86130-163-6

ABMT Band 40
Heier, Silke
Leistungsbewertung der UMTS Funkschnittstelle
1. Auflage 2003, 281 Seiten
ISBN 3-86130-167-9

ABMT Band 41
Schieder, Andreas
Echtzeitdienste in paketvermittelnden Mobilfunknetzen
1. Auflage 2003, 187 Seiten
ISBN 3-86130-168-7

ABMT Band 42
Esseling, Norbert
Ein Relaiskonzept für das hochbitratige drahtlose lokale Netz HIPERLAN/2
1. Auflage 2004, 312 Seiten
ISBN 3-86130-169-5

ABMT Band 43
Farjami, Peyman
Agentenbasierte Dienste in Telekommunikationsnetzen
1. Auflage 2004, 248 Seiten
ISBN 3-86130-170-9

ABMT Band 44
Kriengchaiyapruk, Tham
Dynamic Channel Allocation in UTRA-TDD
1. Auflage 2004, 142 Seiten
ISBN 3-86130-171-7

ABMT Band 45
Sievering, Peter
Dimensionierung und Leistungsbewertung von TETRA-Bündelfunksystemen
1. Auflage 2004, 336 Seiten
ISBN 3-86130-174-1

ABMT Band 46
Wijaya, H.
Broadband Multi-Hop Communication in Homogeneous and Heterogeneous Wireless LAN Networks
1. Auflage 2005, 306 Seiten
ISBN 3-86130-175-X

ABMT Band 47
Seoung-Hoon Oh
Satellite- UMTS - Specification of Protocols and Traffic Performance Analysis
1. Auflage 2005, 188 Seiten
ISBN 3-86130-176-8

ABMT Band 48
Forkel, I.
Performance Evaluation of the UMTS Terrestrial Radio Access Modes
1. Auflage 2005, 3112 Seiten
ISBN 3-86130-177-6

ABMT Band 49
Becker, C.
Leistungsbewertung CEN-DSRC-basierter Mautsysteme
1. Auflage 2005, 202 Seiten
ISBN 3-86130-178-4

ABMT Band 50
Berlemann, L.
Distributed Quality-of-Service Support in Cognitive Radio Networks
1. Auflage 2006
ISBN 3-86130-179-2

ABMT Band 51
Luo, J.
Joint Radio Resource Management for Multi-link Terminals
1. Auflage 2006, 190 Seiten
ISBN 3-86130-930-0

ABMT Band 52
Orfanos, S.
Development and Performance Evaluation of an Adaptive MAC-CDMA Wireless LANs with QoS Support
1. Auflege 2006, 248 Seiten
ISBN 3-86130-931-9

ABMT Band 53
Siebert, M.
Interworking of Wireless and Mobile Networks based on Location Information
1. Auflage 2006, 372 Seiten
ISBN 3-86130-932-7

ABMT Band 54
Zhao, R.
Mesh Distributed Coordination Function for Efficient Wireless Mesh Networks Supporting QoS
1. Auflage 2007, 180 Seiten
ISBN 3-86130-933-5

ABMT Band 55
Gehlen, G.
Mobile Web Service - Concepts, Prototype, and Traffic Performance
1. Auflage 2007, 300 Seiten
ISBN 3-86130-934-3

ABMT Band 56
Radio Resource Control Performance of the Mobile Data Service EGPRS
1. Auflage 2007, 360 Seiten
ISBN 3-86130-935-1

ABMT Band 58
Jelena Mirkovic
Design and Performance Analysis of MIMO Based WLANs
1. Auflage 2009, 190 Seiten
ISBN 3-86130-938-6

ABMT Band 59
Tim Irnich
A New Methodology for Radio SpectrumRequirement Estimation of WirelessCommunication Systems
1. Auflage 2009, 210 Seiten
ISBN 3-86130-939-4

ABMT Band 60
Stephan Göbbels
Smart Caching for Continuous Broadband Services in Intermittent Wireless Networks
ISBN 3-86130-940-8

ABMT Band 61
Michael Einhaus
Dynamic Resource Allocationin OFDMA Systems
1. Auflage 2009, 216 Seiten
ISBN 3-86130-941-6

ABMT Band 62
Erik Paul Weiß
Zur Spektrumseffizienz von Schicht-2 Relais im zellularen Mobilfunk1. Auflage 2010, 240 Seiten
ISBN 3-86130-942-4

ABMT 63
Arif Otyakmaz
Multi-Hop Cellular Radio Network Integrating Frequency and Time Duplexing
1. Auflage 2011, 170 Seiten
ISBN 3-86130-943-2

Aachener Beiträge zur Mobil- und Telekomunikation

ABMT Band 64
Sebastian Max
Capacity and Efficienty of IEEE 802.11n in Wireless Mesh Operation
1. Auflage 2011, 180 Seiten
ISBN 3-86130-944-0

ABMT Band 65
Fahad Aijaz
Mobile Server Platform
1. Auflage 2011, 244 Seiten
ISBN 3-86130-383-3

ABMT Band 66
Zheng Xie
resource Allocation and Reuse for Inter-Cell Interference Mitigation in OFDMA-based Communication Networks
1. Auflage 2012, 230 Seiten
ISBN 3-86130-314-0

ABMT Band 67
Matthias Malkowski
Performance Evaluation of Packet Switched Services in UMTS
1. Auflage 2012, 200 Seiten
ISBN 3-86130-320-5

ABMT Band 68
Guido Hiertz
Medium Access Control in IEEE 802.11. Wireless Mesh Networks
1. Auflage 2012, 212 Seiten
ISBN 3-86130-512-7

ABMT Band 69
Bernhard Walke
ComNets 1990-2012
1. Auflage 2012, 186 Seiten
ISBN 3-86130-511-9

ABMT Band 70
Ralf Jennen
Voice over IP Capacity of IEEE 802.11 WLAN and IEEE 802.21 based Interworking Performance of WLAN and Mobile LTE Networks
1. Auflage 2013, 270 Seiten
ISBN 3-86130-959-8

ABMT 71
Klaus Sambale
Cellular Radio Relay Placement for Optimized Capacity
1. Auflage 2014, 200 Seiten
ISBN 3-86130-475-3

ABMT 72
Benedikt Wolz
Performance Evaluation of Coordinated Beamfoarming in LTE-Advanced Systems
1. Auflage 2015, 186 Seiten
ISBN 978-3-95886-014-8

ABMT 73
Karsten Klagges
VoIP Performance of the Relay-enhanced IEEE 802.16m Wireless Broadband system
1. Auflage 2015, 182 Seiten
ISBN 978-3-95886-022-3

Aachener Beiträge zur Mobil- und Telekomunikation

Aachener Beiträge zur Mobil- und Telekomunikation